SUPERCONDUCTIVITY

The novel properties of superconductors has led to many proposals for the use of superconductors in electronic and electrical engineering applications. The properties of this class of materials, and the means by which these might be enhanced are discussed. The technological uses of bulk superconductors are covered, and many sensitive measuring instruments constructed with superconductors and their advantages over conventional designs are discussed in detail.

Other titles in this series

EE/1 THRESHOLD LOGIC
S L Hurst

EE/2 THE MAGNETRON OSCILLATOR
E Kettlewell

EE/3 MATERIAL FOR THE GUNN EFFECT
J W Orton

EE/4 MICROCIRCUIT LEARNING COMPUTERS
I Aleksander

EE/5 D C CONDUCTION IN THIN FILMS
J G Simmons

EE/6 READING MACHINES
J A Weaver

EE/7 SPEECH SYNTHESIS
J N Holmes

EE/8 GLOW DISCHARGE DISPLAY
G F Weston

Other Electrical Engineering titles available

TL/EE/1 LINEAR ELECTRIC MOTORS
E R Laithwaite

TL/EE/2 AN INTRODUCTION TO THE JOSEPHSON EFFECTS
B W Petley

All published by Mills and Boon Limited

M & B Monographs EE/9

General Editor: J Gordon Cook, PhD, FRIC

Superconductivity

I M Firth, PhD
University of St Andrews

Mills & Boon Limited
London

First published in Great Britain 1972
by Mills & Boon Limited, 17–19 Foley Street,
London, W1A 1DR

ISBN 0.263.05002.5

Set in 11/12 pt. Monotype Baskerville, printed by photolithography, and bound in Great Britain at The Pitman Press, Bath

CONTENTS

1. Introduction

The discovery of superconductivity resulted from experiments carried out by Professor Heike Kamerlingh Onnes of the University of Leiden in 1911. Onnes discovered, when examining the electrical resistance of pure mercury at low temperatures, that the resistivity, instead of decreasing continuously as expected upon cooling, vanished abruptly and completely at 4·15 K. His further work on this topic up to 1913 confirmed this unexpected result and it was at this point that the word superconductivity was used to describe this new state of matter.

In 1908, Kamerlingh Onnes had liquefied the last noble gas helium (boiling point 4·2 K), and discovered that below 2·2 K a peculiar change takes place in the liquid, a change which is now recognised as the superfluid phase of liquid helium. Both superconductivity and superfluidity represent fundamentally new aspects of aggregate matter which have responded in some measure to theoretical scrutiny by quantum mechanics. Our understanding of these co-operative phenomena has thereby been increased, but it is nevertheless unlikely that they would have been predicted from advanced theoretical grounds alone.

Even now, more than 50 years after the discovery of superconductivity, there is no comprehensive theory which enables accurate predictions to be made. There are well developed and self consistent descriptions of superconductivity, but these do not provide a basis on which to make predictions about the properties of superconductors, or even to indicate which materials are likely to be superconductors. Progress in the discovery of materials which show no resistance to the flow of electrons has suffered particularly from this lack of theoretical predictability, and discovery has often been the result of an empirical approach.

2. Basic Properties

Superconductivity is a co-operative phenomenon of electrical conductivity. It is displayed when the normal processes of scattering between electrons and lattice which gives rise to electrical resistance are removed; a superconductor shows no impediment to the flow of a steady electrical current. Like all co-operative phenomena the state of superconductivity cannot be present above a certain critical temperature, T_c, or transition temperature (Fig. 1). As it is an electromagnetic phenomenon superconductivity can also be destroyed by imposing a magnetic

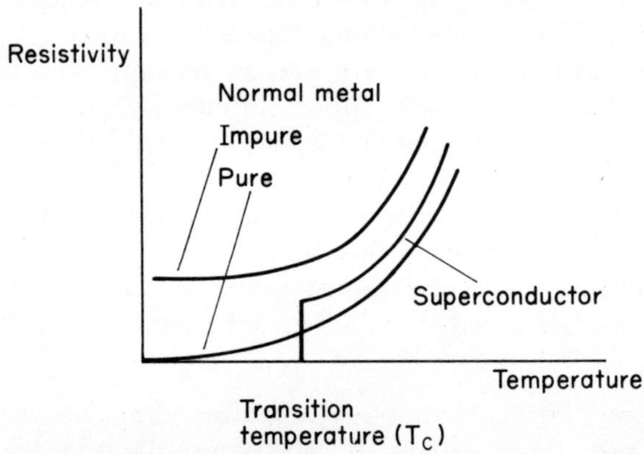

Fig. 1 Resistance of a superconductor vanishes completely at the transition point

field above a certain critical field, H_c. The magnetic field may be applied externally or may arise from an electrical current passing through the superconductor, so that the current a superconductor can sustain is limited also to less than a critical current, J_c. A superconductor occurs therefore within a surface passing through T_c, H_c and J_c in Fig. 2, and outside this surface the superconductor reverts to its normal state of electrical resistance.

The critical temperature of all known superconducting materials is in the cryogenic temperature range (liquid helium is used as a refrigerant) and the highest critical temperature so far discovered for any material is 20·8 K.

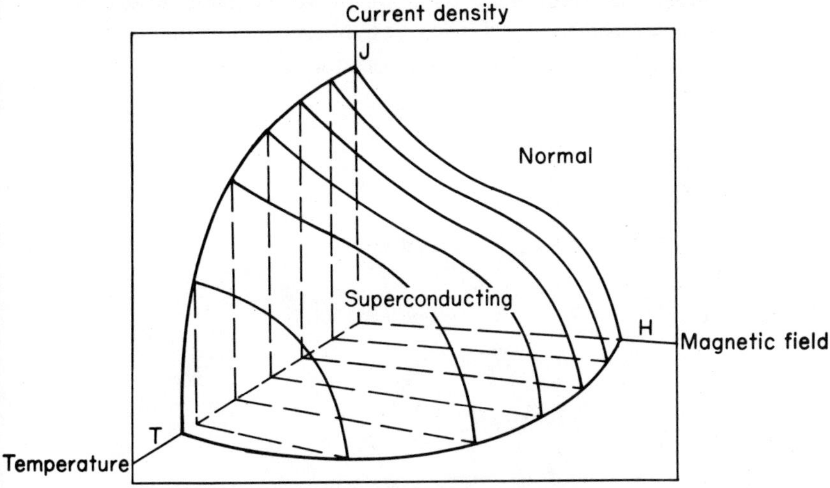

Fig. 2 Surface which separates normal from superconducting states in Type I materials. This surface is the critical surface

The persistent electrical currents which can be established in electromagnets made from superconducting materials as a consequence of their zero resistance are one of the most dramatic manifestations of superconductivity. One of the more interesting properties of a superconductor, however, is it diamagnetism, or impermeability to magnetic field. When cooled in a magnetic field less than the critical value a superconductor expels magnetic flux when the critical temperature is reached. The way in which this process takes place divides superconducting materials into two types. In Type I materials the magnetic field is totally excluded from the interior up to H_c, and above H_c the magnetic field penetrates fully into the sample and normal electrical resistance is restored. In Type II materials, while the perfect superconducting state exists for fields less than a lower critical magnetic field, H_{c1}, partial penetration of magnetic field occurs and increases up to a higher critical field at which penetration is complete and normal electrical resistance is finally restored. The behaviour of Type II materials is described by Fig. 3, the state of zero resistance occurring up to a surface passing through the points T_c, H_{c2}, J_{c2}. The

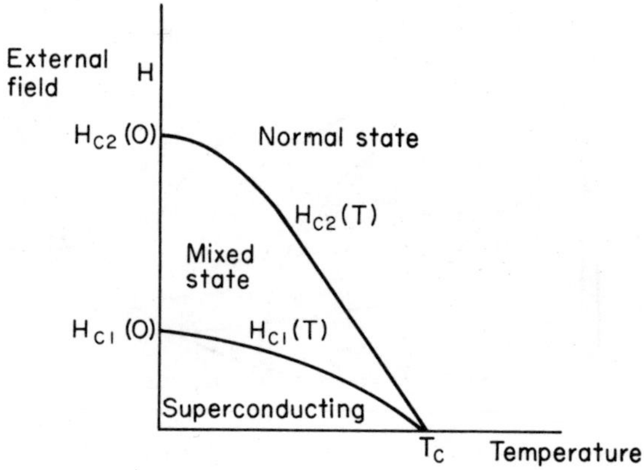

Fig. 3 Flux penetration occurs in the mixed state in Type II superconductors

magnetic behaviour of the two types described by Fig. 3 in which the magnetisation (diamagnetism) is plotted against magnetic field: Fig. 4 shows reversible behaviour which is not achieved in practice in many materials (see Fig. 14). Type I materials are restricted to those which show low values of T_c and H_c, whereas materials which are of interest in technological projects are invariably of Type II, and in particular those which exhibit high field superconductivity (see Table I).

Whereas the number of materials discovered to be superconducting has grown continuously and rapidly since 1911, the maximum transition temperature attainable

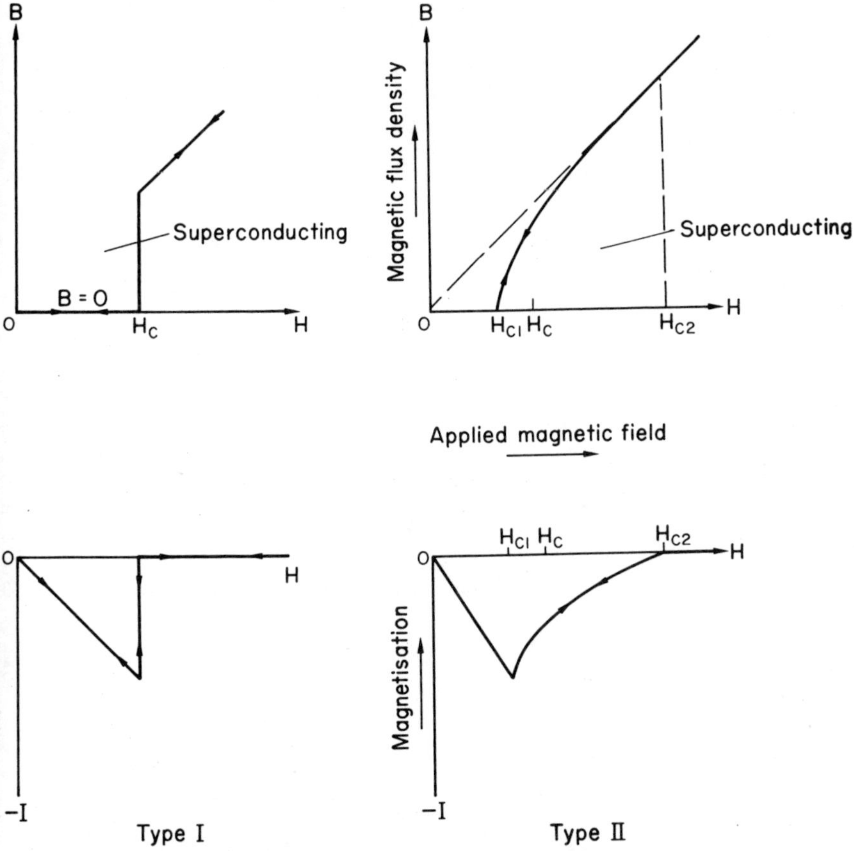

Fig. 4 Magnetisation and magnetic flux density of a superconductor: Type I and Type II

has remained always in the very low temperature range (Fig. 5). The discovery in 1959 that niobium-tin (Nb_3Sn) is superconducting up to 18 K followed other discoveries at nearby temperatures at about the same time. But between 1961 and 1968 no further increase was reported and 18 K began to appear as a barrier. Since then this temperature has been broken by two materials with higher transition temperatures: $Nb_3(Al_{0\cdot8}\,Ge_{0\cdot2})$ at

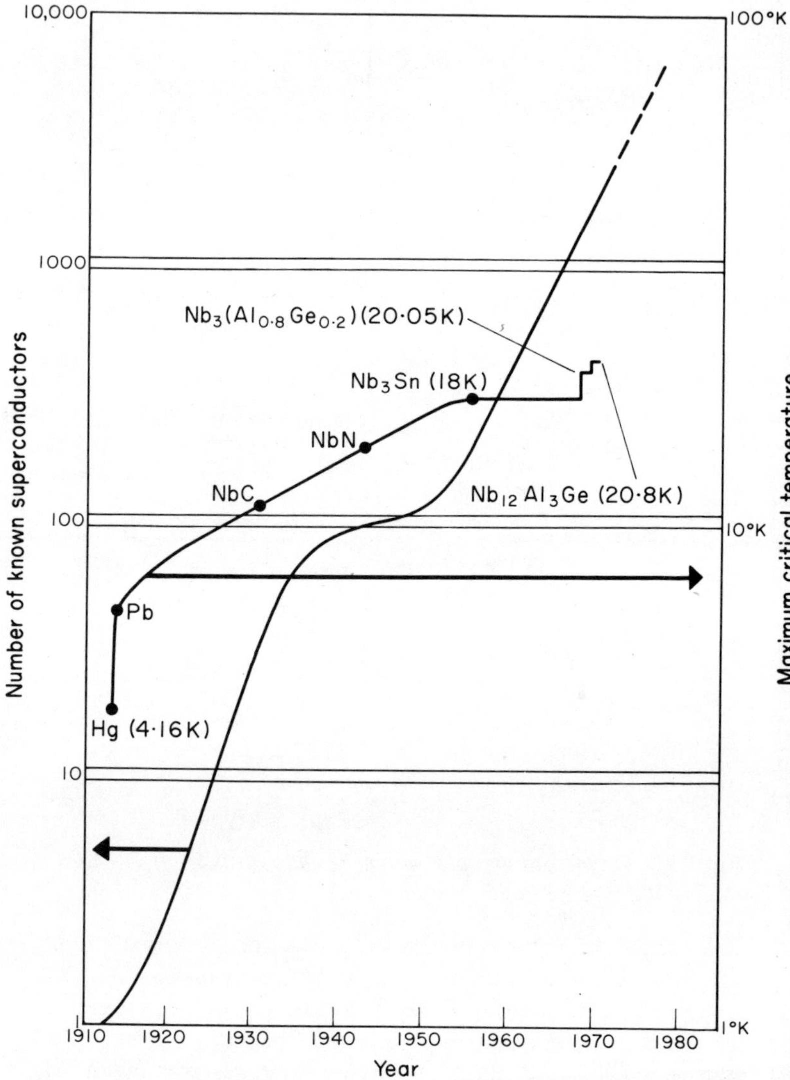

Fig. 5 Growth of the number of superconductors and in the maximum critical temperature

20·05 K and $Nb_{12}Al_3Ge$ at 20·8 K. These materials show, therefore, superconducting properties close to the boiling point of hydrogen (20·3 K); this fact has technological importance in refrigeration, as hydrogen is both a cheaper and more efficient coolant than helium.

Table I

Superconducting properties of some materials

MATERIAL	T_c, K	H_{c1}, or H_c	H_{c2}
Ir	0·14	19 Gauss	extrapolated to 0 K
Cd	0·56	29 G	,,
Zn	0·85	53 G	,,
Al	1·19	99 G	,,
V	5·3	1020 G	,,
Pb	7·2	803 G	,,
Nb	9·2	1·1 KG	5·5 KG at 4·2 K
Nb_5Ti_5	9·5		120 KG ,,
Nb_2Zr	10·8		80 KG ,,
NNb	15·6		9·5 KG at 13·2 K
Si V_3	16·9		156 KG extrapolated to 0 K
Nb_3Sn	18·05	190 G	220 KG at 4·2 K
$Nb_3Al_{0·8}Ge_{0·2}$	20·2		
$Nb_{12}Al_3Ge$	20·7		400 KG at 0 K

BEHAVIOUR IN MAGNETIC FIELDS

Some insight into the behaviour of superconductors can be obtained by considering their macroscopic properties. The central phenomena are the complete loss of electrical resistance below a critical temperature (Fig. 1) and, while in the superconducting state, the exclusion of magnetic flux (Fig. 4). The exclusion of magnetic flux is total in the region of ideal behaviour up to H_c in the case of Type I, and H_{c1} in Type II materials. In Type II materials another type of behaviour, known as the mixed state, allows the superconductive state to exist to larger fields, up to H_{c2}; this is a state pictured as interwoven superconducting and normal regions, in which magnetic flux penetrates the material being contained in the normal regions.

In the region of ideal behaviour the superconductor acts as a perfect diamagnetic material. A solid material cooled below its transition temperature excludes an externally applied magnetic field whether this is applied to the sample during cooling through the transition temperature, or afterwards so long as the field is less than critical. The application of magnetic field induces lossless currents which circulate on the surface of the specimen in a manner which creates a magnetic induction, B, everywhere equal and opposite to the applied field flux density, B_e, thereby producing complete cancellation of the magnetic induction, inside the sample (Fig. 6).

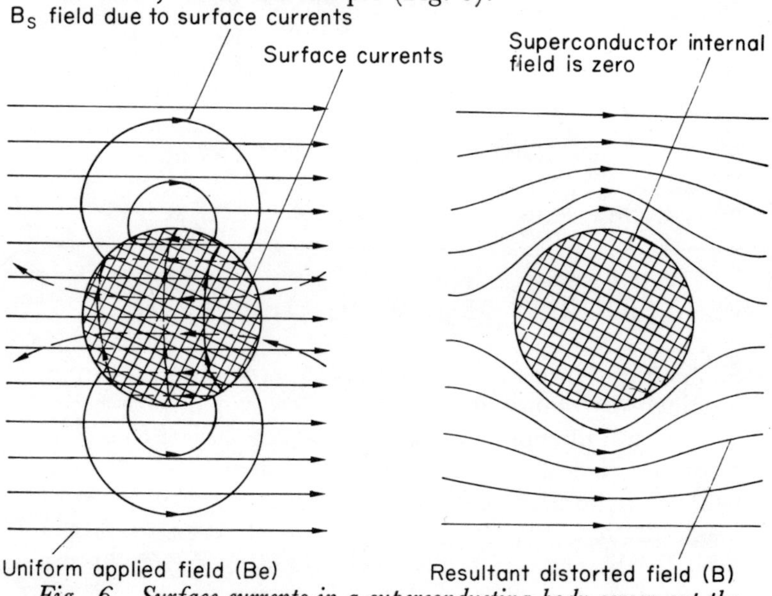

Fig. 6 Surface currents in a superconducting body screen out the externally applied field

Inside the superconductor the magnetic induction is

$$B = B_e - B_s = 0$$

This effect occurs for fields less than critical and is reflected in a negative magnetisation of the specimen shown in Fig. 4.

The superposition of the field produced by these screening currents and the applied field produces a perturbed external field distribution (Fig. 6). In the equatorial region the magnetic field is greater than the externally applied field. This increase in magnetic field occurs in all specimen shapes except a long rod; in this case, the screening currents behave like a long solenoid with no external magnetic field.

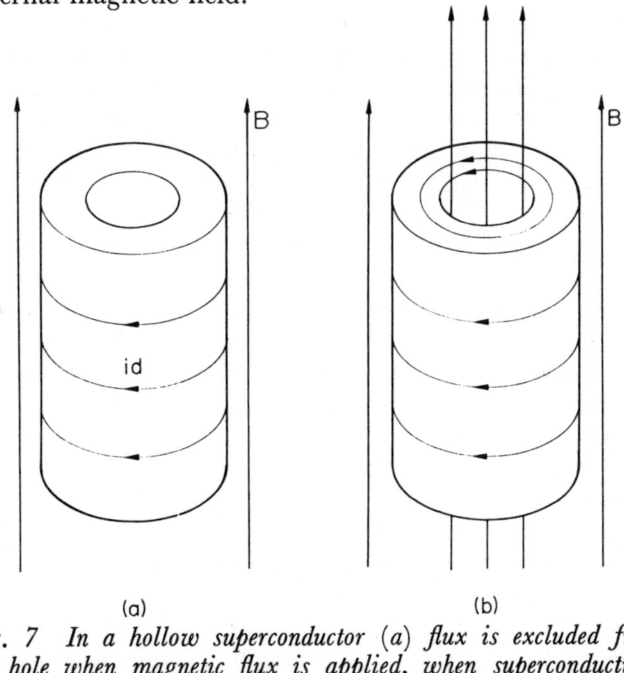

Fig. 7 In a hollow superconductor (a) flux is excluded from the hole when magnetic flux is applied, when superconducting. (b) Flux penetrates hole when cooled below T_c in a magnetic field. Circulating surface currents are shown

The situation is different, however, if the superconducting specimen has a hole through it. When a magnetic field, below the critical value, is applied after the sample has been cooled below T_c diamagnetic screening currents, i_d, circulate over the outside surface. The magnetic induction produced by these currents cancels the magnetic induction through the hole also, so that no magnetic field passes through it (Fig. 7(*a*)). However, should the specimen be cooled through its transition temperature in the presence of an applied field, when the resistance vanishes circulating currents must appear to expel magnetic flux from the interior of the specimen. The magnetic field which passes through the hole when it is above the critical temperature remains undisturbed, the currents maintaining the differences in magnetic induction (Fig. 7(*b*)). Currents on the inside surface of the hole flow counter to those on the outside. As there is no means of dissipating the energy associated with the inside current, this current continues to maintain the magnetic field within the hole in the superconductor for an indefinite time, even if the external field is altered (always less than H_c) or reduced to zero. Such persistent currents are employed to advantage in coils made from superconducting wire. Once magnetic flux has been established within the coil, say by passing current from an external circuit, the low temperature coil circuit is completed by switching in a superconducting link, the current passing through the completed coil with link becoming persistent even although the battery is removed. Large stable magnetic fields can be maintained indefinitely without the continuous dissipation of electrical energy.

BEHAVIOUR OF TYPE I MATERIALS

A great deal of what is understood theoretically about the superconducting state has come from investigations on the destruction of this state and the reversion of materials to the normal state of electrical resistance. In Type I materials this transition occurs at a particular critical value of an applied field, H_c, but the transition is sharp only when the superconductor is in the form of a long cylinder with its axis parallel to the applied field. For all other sample shapes or orientations the transition begins at a magnetic field less than critical and proceeds gradually until at H_c it is complete, normal resistance only then being fully restored.

In the general case, a superconductor begins to lose its zero resistance when the total magnetic field at any point on its surface exceeds the critical value H_c. The field may be due to the field created by a transport current together with an externally applied field. The value of the critical current depends, therefore, on the applied field, and also on the shape of the sample and its orientation in the applied field. The critical current is defined most generally as the maximum current flowing before a measurable voltage across the sample, i.e. resistance, is first noted.

For example, a superconducting sphere experiences a larger magnetic field at its equator than at its poles (Fig. 6). As the equatorial field is 1·5 times the external field, when the external field is only $\frac{2}{3}H_c$ the equatorial field has the value of H_c. The reason for this is that the superconducting screening currents which create the negative magnetisation of the specimen in excluding the magnetic field from the material enhance the external field around the equator. It might be supposed that at this point penetration of field occurs uniformly in an annulus around the sphere, but if this were so the equatorial field would fall below H_c for as the material of the annulus becomes normal its magnetisation becomes zero. This paradox of having a completely normal sample in a field below critical is resolved if an intermediate state is allowed between the external field value at which the surface field first becomes critical, and the critical value H_c.

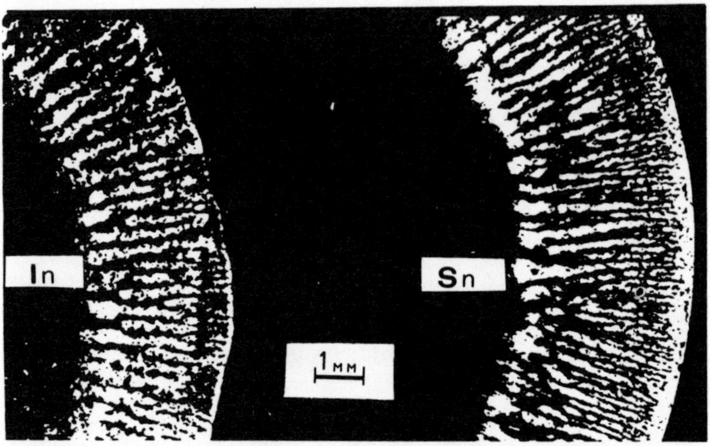

Fig. 8 Flux penetration in Type I materials from edge of disc. Dark—superconducting, light—normal material. (Courtesy J. F Allen and R. Olafsson)

This intermediate state contains alternate lamellae of normal and superconducting material with surfaces parallel to the field direction and thicknesses so that all the penetrating field is carried by the normal lamellae. As the external magnetic field is increased the superconducting lamellae contract, and at the outer edges the normal regions coalesce. Eventually the centre of the sample begins to fill up with normal regions which are often nucleated from the peripheral normal regions (Fig. 8). The magnetic behaviour of the intermediate state is shown in Fig. 9. The magnetisation is not reversible, a lower field being required to induce all normal regions to leave the interior of the specimen as the field is lowered.

The formation of intermixed normal and superconducting regions occurs because of the presence of a surface energy between the two regions. This surface energy is an additional contribution to the Gibb's free energy which is proportional to the total surface area of the boundary between the two phases. In Type I superconductors this surface energy is positive, so that the free energy is minimised when the boundary area is small. In this way the lamellae are found to be few and are thick (Fig. 8).

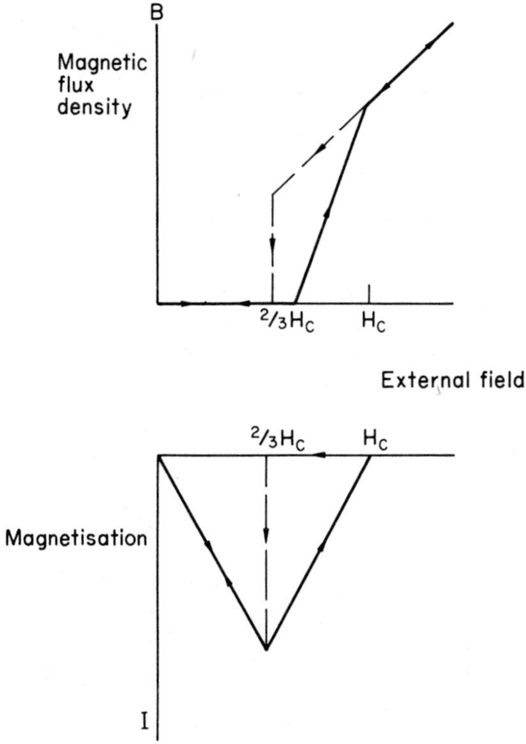

Fig. 9 Magnetic behaviour of superconducting sphere in external field

The origin of the surface energy at a boundary between normal and superconducting regions can be explained by considering how the magnetic flux density, B, and the number of electrons which show superconducting properties change across the boundary. Firstly, electromagnetic theory requires that the change from a uniform magnetic field outside a superconductor (in a normal region) to zero field inside the superconductor cannot be perfectly sharp but must take place over a finite width. The field penetrates a short distance into a superconducting region, called the London penetration depth λ, such that

$$\lambda^2 = mc^2/4\pi n_s e^2$$

where m and e are the electronic mass and charge, c is the velocity of light and n_s is the number of electrons per cubic centimetre which show superconducting properties. It is within this depth λ that superconducting screening currents flow to exclude the magnetic field which decreases within the superconductor according to

$$H = H_0\, e^{-x/\lambda} \quad .$$

λ is of the order of 10^{-5} cm and is strongly dependent on temperature, and also to an extent on sample shape and size.

Secondly, the number of superconducting electrons, n_s, which show a high degree of order in this co-operative phase can only change appreciably within a distance called the coherence length ξ. This distance is of the order of 10^{-4} cm in pure Type I materials, but is smaller in samples which contain impurities. Evidence for a coherence length comes from the sharpness of the transition at the critical temperature, often less than 10^{-5} K wide, an observation which suggests that the ordering of electrons is high and changes little over lengths large in comparison with the electronic mean free path. Coherence is a feature of the microscopic theory of superconductivity.

By means of these two characteristic lengths the origins of a surface energy can be simply explained. For the boundary to be in equilibrium, the Gibb's free energy on each side must be the same. Considering the change in energy in going from the normal to superconducting side of the boundary (Fig. 10) the energy of the superconducting phase is lowered because of the presence of ordered superconducting electrons. On the other hand, because the superconductor has a magnetisation which cancels the magnetic flux density inside there is a positive contribution to the free energy. The first of these contributions to the free energy changes with the coherence length ξ and the second with the penetration depth λ; the net difference about the boundary, which is positive, is approximately $\frac{1}{2}\mu_0 H_c^2(\xi - \lambda)$ per unit area (Fig. 10).

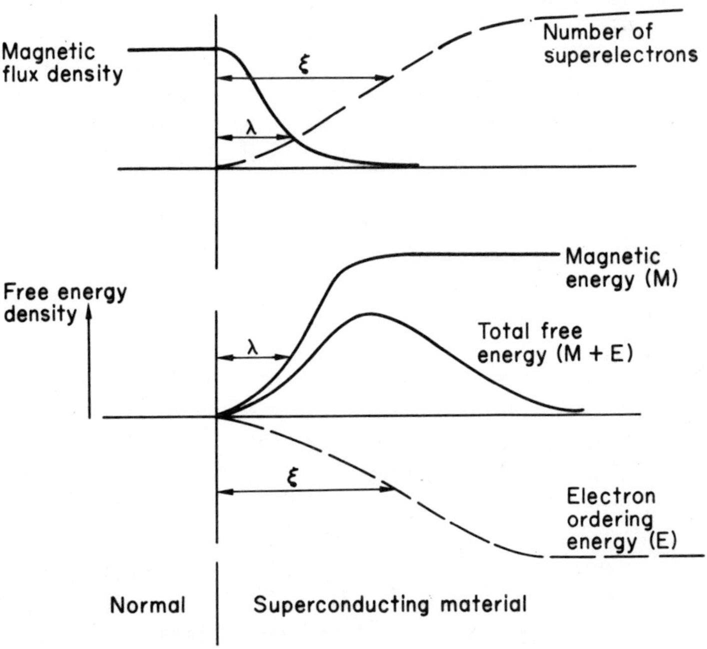

Fig. 10 Positive surface energy arises in Type I superconductors because coherence length ξ > penetration depth λ

BEHAVIOUR OF TYPE II MATERIALS

The behaviour of the other type of superconductor, Type II, can be equally described in terms of a surface energy. This class of superconductor, which is of technological interest, shows a negative surface energy between superconducting and normal regions. In this material the appearance of a normal region lowers the Gibb's free energy so that it may be energetically favourable during transition for the material to split up into a large number of small normal filaments to produce a maximum boundary area relative to the volume of the normal material. The transition in Type II materials from the superconducting to the normal state takes on a different form to that of Type I, or perfect superconductivity, as seen in Fig. 4. The expulsion of an external magnetic field is complete up to the first critical field H_{c1}, but above this magnetic field penetrates the specimen in filaments of normal material each carrying a quantum unit of magnetic flux, lying parallel to the imposed field. The number of filaments increases with applied field, but the material shows no electrical resistance until an upper critical field, H_{c2}, is reached at which point the filaments occupy the whole of the specimen.

The filaments are surrounded by superconducting screening currents and repel one another so that they occupy the whole specimen volume and form a regular array. That this array is regular and of triangular form in accordance with Abrikosov's predictions has been shown recently by Essman and Trauble using an electron microscope (Fig. 11). Each filament contains a quantum of magnetic flux, ($h/2e = 2{\cdot}07\ 10^{-11}$ Tesla, where h is Planck's constant) and their presence between H_{c1} and H_{c2} is referred to as the mixed state. The penetration of flux in the mixed state occurs in Type II superconductors no matter what the shape of the specimen may be, but the intermediate state occurs in Type I materials with its concomitant of flux penetration only in samples which have a demagnetising factor which is not zero—that is, in any sample which is not a long thin wire along the field direction.

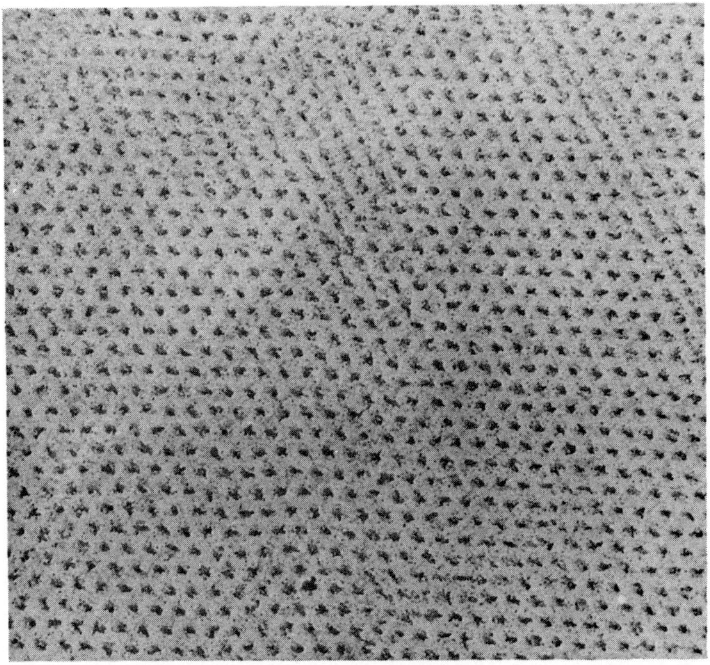

Fig. 11 Direct observation of individual flux lines in the mixed state of Type II superconductors. The flux lines are arranged in a regular Δ *lattice* (Pb *with 6·3 at per cent* In, *Mag. 9,000*×) (*Courtesy U. Essman*)

The magnetic surface energy associated with Type II materials can likewise be explained in terms of the penetration depth λ and coherence length ξ. If the coherence length is shorter than the penetration depth, the surface energy is still approximately $\frac{1}{2}\mu_0 H^2(\xi - \lambda)$ per unit area as in the intermediate state, but is now negative (Fig. 12). As ξ depends upon the mean free path of electrons and is consequently decreased by impurities and mechanical strain, it can easily become less than the penetration depth in impure metals and alloys. Such materials are usually Type II superconductors, whereas most pure metals are Type I.

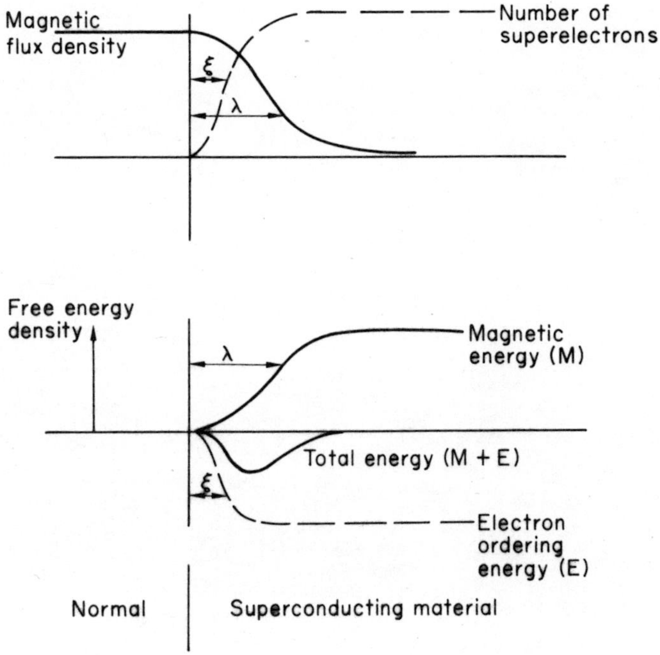

Fig. 12 Negative surface energy arises in Type II superconductors because coherence length $\xi <$ penetrates length λ

The parameter that determines the class into which a particular material falls is the ratio K of penetration depth λ to coherence lengths ξ,

$$K = \frac{\lambda,}{\xi}$$

the Ginsburg-Landau constant. A detailed treatment shows that the value of K which separates the two types is $1/\sqrt{2}$ so that

$K < 0{\cdot}71$ Type I, positive surface energy, and
$K > 0{\cdot}71$ Type II, negative surface energy.

As K increases above 0·71 the upper critical field increases, but H_{c1} the point of first entry of field into the superconductor decreases (Fig. 13). Of the pure metals, only niobium (K = 0·78), vanadium (K = 0·85) and technetium (K = 0·92) are intrinsic Type II superconductors. However, most alloys are Type II, and it is alloys of particular compositions which are found to have the highest critical fields (H_{c2}), that is, K is large (Table I).

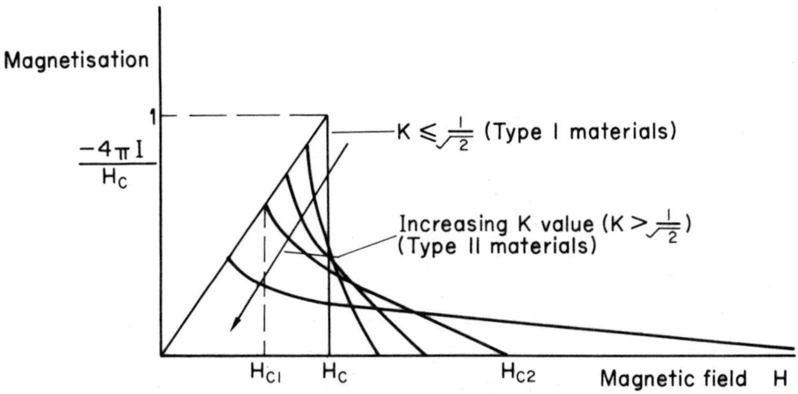

Fig. 13 Change from Type I to Type II with changing Ginsburg-Landau constant, K

In all but very pure, fully annealed and carefully prepared specimens the magnetisation curve (Fig. 13) shows hysteresis. Magnetic hysteresis of Type II superconductors is caused by the presence of impurities, or by structural defects such as dislocations, precipitates of other phases, gain boundaries and voids, in the material. In some samples the amount of hysteresis may be large, as shown diagrammatically in Fig. 14. Such hysteretic effects may be unwanted in situations where the magnetic field is changing, but the fact that there is strong interaction between defects and magnetic flux in the mixed state has the useful feature that filaments of flux can be pinned to defects with a consequent restriction on their movement within the material. This situation is highly important to the technological use of Type II superconductors in producing high magnetic fields.

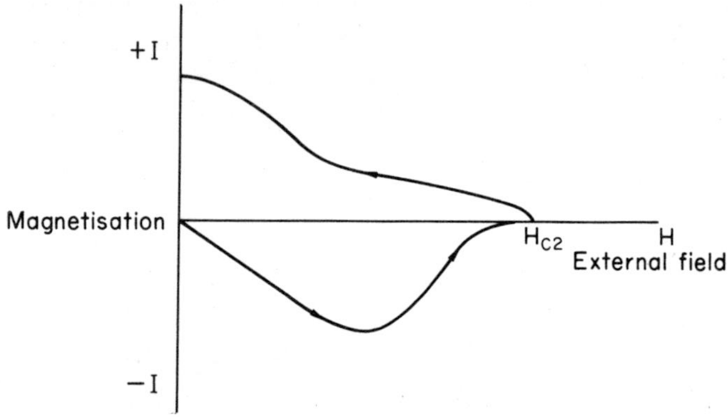

Fig. 14 Hysteresis in magnetisation for superconductor containing defects—cold worked sample

THE MIXED STATE

In the mixed state, above H_{c1}, although transport current is able to flow without resistance through the bulk superconductor which is threaded by the array of filamentary flux lines, there is nevertheless an interaction between the transport current and these flux lines. This interaction gives rise to resistance and power dissipation in the material. Where a superconductor is transverse to a magnetic field, the usual situation where a magnetic field is created by a superconducting coil, the filaments of flux penetrate across the direction in which the transport current is flowing. A Lorentz force is exerted between the flux threading the filaments of normal material, and the superconducting transport current. The filaments experience of a force at right angles to themselves and the transport current, and tend to move across the sample. Their motion is restricted by the pinning force provided by defects. Not every filament of flux is directly pinned to some defect, but the repulsion between the penetrating filaments gives the whole matrix of filaments a certain rigidity, so that if some filaments are pinned the whole matrix is stationary. So long as the Lorentz force on a filament is less than the average pinning force of a filament no motion takes place.

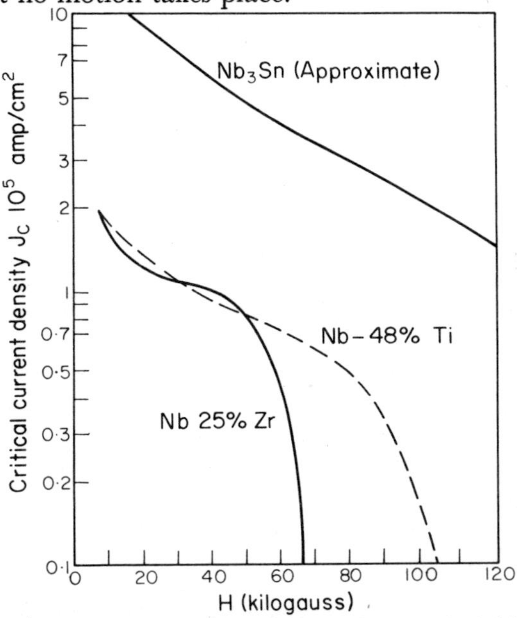

Fig. 15 Critical current as a function of magnetic field for three commonly used superconducting alloys

As the transport current is increased the filaments are dislodged from their pinning sites and move across the specimen. Work is required to maintain their motion and this can only be supplied by the transport current. Such motion of the penetrating filaments of flux is reflected, therefore, as an energy dissipation by the transport current; in other words a voltage appears across the sample, and an effective resistance in the superconductor appears. The critical current, defined as the maximum current passed before a voltage appears, is therefore higher in materials containing many defects (of any type) which increase the average pinning force on the filamentary flux lines which penetrate the material in the mixed state. Typical curves showing the critical current for well known superconducting materials are shown in Fig. 15. The critical current becomes very small as H_{c2} is approached.

Two features of this dissipation process are of great importance in the practical use of superconductors. During the build up of current in a superconductor, for instance a superconducting solenoid, flux begins to penetrate as the field is increased above H_{c1}, so that dissipative forces are present during the process of reaching an operation point. Although the heat evolved in this process may be small, so are the specific heats of materials at the temperatures of liquid helium. Consequently a significant local rise in temperature may be produced, and more flux released. The design of the superconducting assembly must prevent large local rises in temperature otherwise material may be driven above its critical temperature, whereupon resistive heating would appear leading to a catastrophic re-establishment of the normal state.

Secondly, even when a steady current is flowing, filaments of flux may jump from one pinning centre to another, thereby producing local heating. The suppression, or evasion, of flux jumping is the essence of stable operation of any Type II superconducting device operating in the mixed state (H_{c1} to H_{c2}). Copper clad cables are usually employed in magnet construction to overcome these difficulties and ensure stable operation. The high thermal conductivity of pure copper prevents the creation of local hot spots and its high electrical conductivity slows down flux motion. Further, in any region of the superconductor which becomes normal, it acts as a protective current path—a shunt—without dissipating too much heat.

3. Microscopic Theory

The most fruitful approach to a microscopic theory has been due to Bardeen, Cooper and Schrieffer* in 1957. This approach, the BSC theory, was to consider the ordering of electrons in the superconducting state as caused by an attractive interaction between pairs of electrons which results from the interaction of electrons with the lattice of the material. This attraction overcomes the natural coulomb repulsion between electrons carried by a current. The clue to this interaction is the experimental fact that the transition temperature and the critical field depend on the atomic mass (isotope effect), and that superconductivity has not been discovered in the best conductors, copper and silver, whereas it is common among the poorer conductors. This suggests that the interaction which produces an ordering is concerned with the vibrations of the lattice of the material.

The attraction suggested by the BCS theory is formed indirectly through the lattice vibrations, or phonons. An incoming electron to a region in the material emits a virtual phonon which in turn is absorbed by another electron, the first thereby changing the momentum of the second electron (Fig. 16). Under appropriate circumstances the force between the two electrons can become attractive so that the electrons form bound pairs. In forming the pair there is a lowering of energy so that between individual free electron-states (normal electrons) and the superconducting pair there is an energy gap; the formation of pairs is thus the favoured state.

In the BCS theory pairs of electrons with equal but opposite momentum with opposite orientation of electron

* Bardeen, J., Cooper L. N., and Schrieffer, J. R. *Physical Review*, **106,** p. 162 (1957).

spin are bound together. The ordering is in electron momentum, not in their position. The attraction between a pair is weak (binding energy = energy gap = $3{\cdot}5\,kT_c$, where k is Boltzmann's constant and T_c is the critical temperature <20 K) which implies that the two electrons are far apart. They are separated by a distance (the coherence length ξ) many thousands of times greater than the distance between lattice ions, and consequently each bound pair ranges over a volume which contains millions of other electron pairs. Overlapping of pairs in this way has important repercussions that arise from the quantum mechanical Pauli exclusion principle, which states that two electrons of the same spin cannot occupy the same state in the lattice. To meet this restriction the momentum of all electron pairs must be correlated—the momentum of the centres of mass of all pairs is the same and in an isolated specimen is zero. Electron pairs are locked together therefore in momentum, so that disturbance of one pair is prevented by scattering by other electron pairs, thereby maintaining the uniformity of momentum.

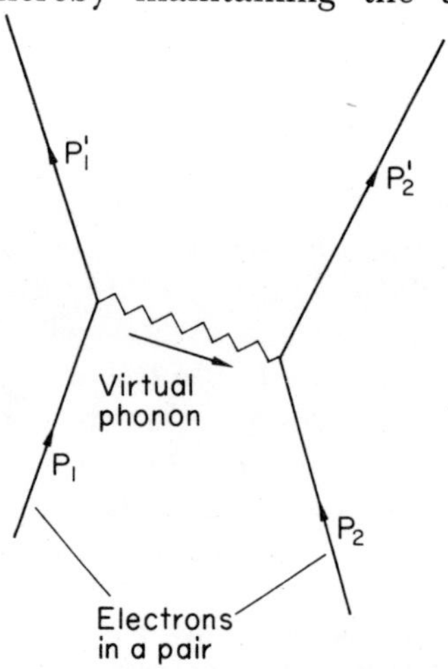

Fig. 16 Schematic representation of attractive pairing mechanism between electrons through interaction with lattice phonons (vibrations)

When a transport current is passed through a superconductor the momentum in the direction of the current increases so that the momentum carried by all pairs is the same. When a current is carried by an ordinary conductor, resistance to current flow is present because the charge carriers can be scattered by impurities, such as lattice defects or thermal vibrations, with a change in momentum of the free electrons. The acceleration of the carriers in the direction of the current is, therefore, impeded. In the superconductor, since the total momentum remains constant, there is no change in the current flowing, no acceleration of charge and consequently no electrical resistance. Resistance can only appear when pairs are broken, a scattering process which requires at least an energy equal to the binding energy, energy which must be supplied from somewhere (increased temperature, etc). The number of paired electrons changes rapidly with increasing temperature until at the critical temperature all electrons are unpaired and the material is in the normal resistive state.

It is useful to consider the superconducting state from the standpoint of electron wavefunctions in elementary quantum mechanics. A pair with total momentum, p, can be represented as a wavefunction which has a de Broglie wavelength of h/p. The wavelength is the same for all pairs of electrons, and because of the overlapping of the wavefunctions of pairs in space the phases of the individual wavefunctions must be the same. Although a wavefunction of an individual pair has a spatial extent of ξ, the coherence length, the phase coherence of the travelling wave representing the overall motion extends over an indefinite distance. In the case of a persistent current in a solenoid wound with superconducting wire, the coherence in phase is over a few miles.

The long-range coherence of the electronic wavefunction in a superconductor makes it generally impossible to introduce any phase differences. When phase differences are present in the wavefunction in a bulk sample it means generally that superconductivity has been destroyed. Such phase differences can be imposed by a sufficiently large current or applied magnetic field which through their potentials can alter quantum mechanically the phase of the electron wave. Nevertheless it has been recently shown that a thin barrier of insulating material interposed between two superconductors gives a means of altering the phase between two regions in a superconducting sample. This idea, introduced by Josephson, has important technological implications which will be discussed later.

THE OCCURRENCE OF SUPERCONDUCTIVITY

The BCS microscopic theory gives a careful and self-consistent description of superconductivity, being confirmed overall by measurements. Nevertheless, it has not been possible to calculate parameters of superconductivity for a given material from first principles. In the search for higher transition temperature the BCS and other microscopic theories have not been able to predict what materials or cystalline structures would be most likely to bear scrutiny. On the other hand simple, partly empirical relations have contributed significantly to the discovery of new high temperature superconductors.

Figure 17 provides a rough survey of the occurrence of superconductivity among the elements in the periodic table. In the case of the non-transition elements the interaction between electrons and lattice vibrations is clear cut, for a consistent isotope effect is found,

$$T_c \propto \frac{1}{M^{\frac{1}{2}}}$$

where M is the atomic mass, and is according to the law of frequency versus mass of a harmonic oscillator.

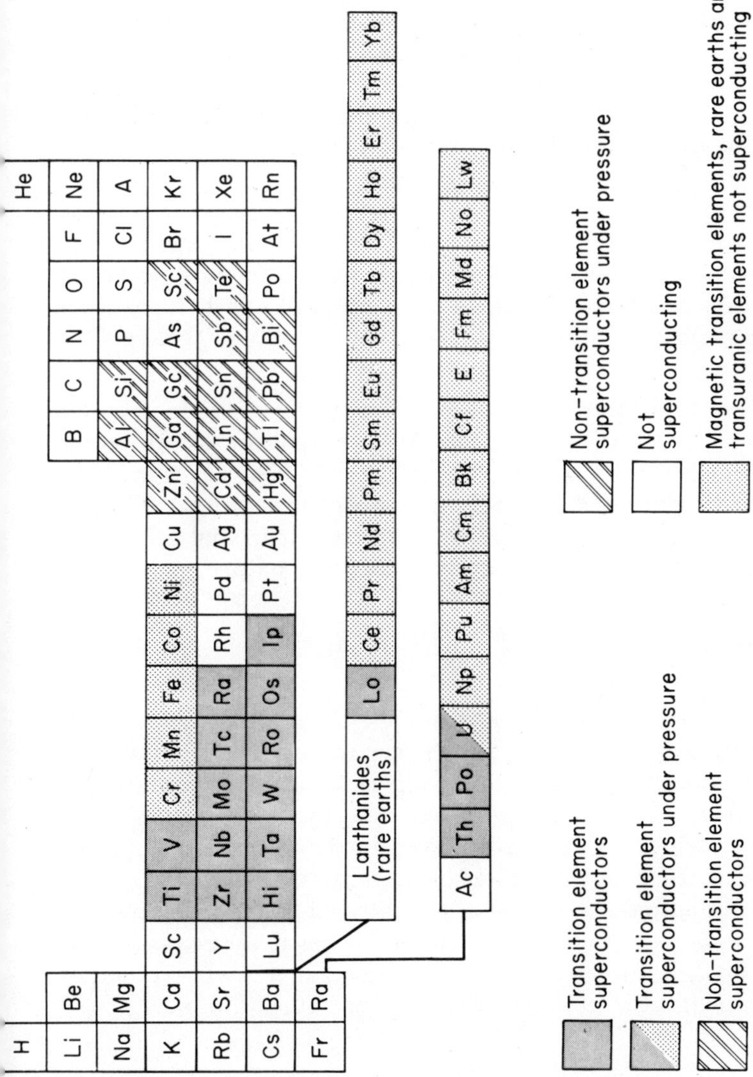

Fig. 17 Occurrence of superconductivity in the elements of the periodic table. (After Matthias)

Hence the interaction between electrons and the lattice vibrations (BCS theory) does cause superconductivity.

In the case of the transition elements (elements with filled electronic shells) this isotopic mass dependence has not been found: for Zr and Ra the mass exponent is zero: for Mo it is $\frac{1}{3}$. From this evidence it would appear that a mechanism other than that of the BCS type for superconduction might be present. Further support for other mechanisms which lead to superconductivity is given by the fact that the transition temperature of non-transition elements never exceeds 10 K, whereas the presence of transition elements in compounds or alloys leads to the high critical temperatures which have been measured.

An important step in establishing the criteria for the superconductive state has been taken by Matthias on the basis of empirical considerations.* The variation of the superconductive transition temperature due to changes in the number of valence electrons can be accounted for by a function of purely qualitative nature,

$$T_c = C\frac{V^x}{M}\mathrm{T(n)}$$

where V is the atomic volume, M the atomic mass, C a constant depending upon crystal structure and x a number. For transition group elements, compounds and alloys T(n), the function describing the qualitative behaviour of T_c, high transition temperatures occur close to 5 and 7 valence electrons per atom (Fig. 18). Compounds with these numbers of valence atoms are most likely to show the highest transition temperatures; this is borne out by all recent experiments in compounds showing the highest critical temperatures around 20 K.

* See Reading List, Review Article 2.

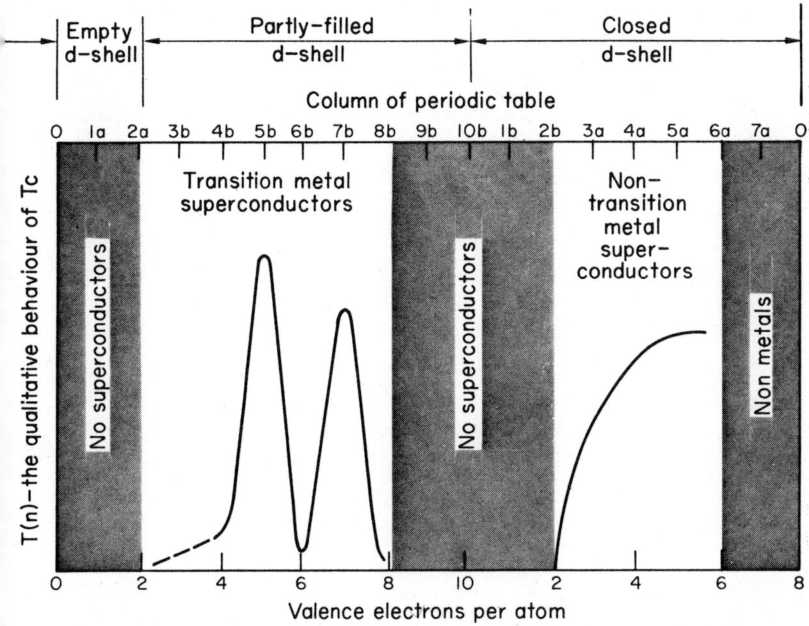

Fig. 18 Maximum values of T(n) *occur close to 5 and 7 valence electrons per atom*

Matthias suggests that the mechanism for superconductivity may not be the same in the transition and non-transition groups. In the BCS microscopic theory the conduction electrons interact with the lattice vibrations, so that the more the lattice vibrates the better. Hence, higher transition temperatures will appear from lower isotopic masses, or lower Debye temperatures. On the other hand, if the mechanism involves the interaction of the atomic electron shells the interaction will be stronger for high atomic mass, as such atoms are more easily polarised. The effect of mass is therefore opposite in each suggested mechanism, and this may be the explanation for the differing isotope effects which are found in superconductivity for different regions of the periodic table, and for the occurrence of very high transition temperatures which are found only with the transition elements and their compounds.

Uranium is a transition element which shows the peculiar behaviour of becoming superconducting only under pressure. Under pressure, the f-electron configuration of uranium changes so that the number of states for electrons to occupy at the Fermi level is altered. The effect of pressure on simple conductivity is not restricted to uranium: hydrostatic pressure has been found to decrease the critical temperature of many elements, including for example barium, bismuth, arsenic and gallium. Investigations on these materials indicates that hydrostatic pressure decreases the number of states available to electrons at the Fermi level. The interaction between electrons and lattice phonons, which gives rise to superconductivity according to the BCS theory, and which can only take place near the Fermi surface, is thereby diminished with a consequent decrease in the superconducting transition temperature. Research into high pressure superconducting properties has led to a greater understanding of the modified crystal forms of materials which can be produced only under pressure.

Considerable attention has been paid to the dependence of critical temperature on various types of crystal structure. The highest critical temperatures so far discovered seem to be confined to the "beta-tungsten" and "alpha-manganese" lattices. The compound of niobium, aluminium and germanium which has shown the highest critical temperature so far discovered, 20·8 K, has a beta-tungsten structure as well as Nb_3Sn (18·75 K) which is used to form superconducting cables. In the alpha-manganese structure are compounds such as Re_3M_0 (9·8 K), Re_3Ta (6·8 K).

Correlations between the critical temperature and other crystal parameters have been shown to exist empirically.

Phillips* has argued that theoretical analysis of structural properties does reveal many coincidences between anomalies in these properties and high values of T_c. The factor which is found to correlate best with high values of T_c is not, Phillips argues, the electron-atom ratio

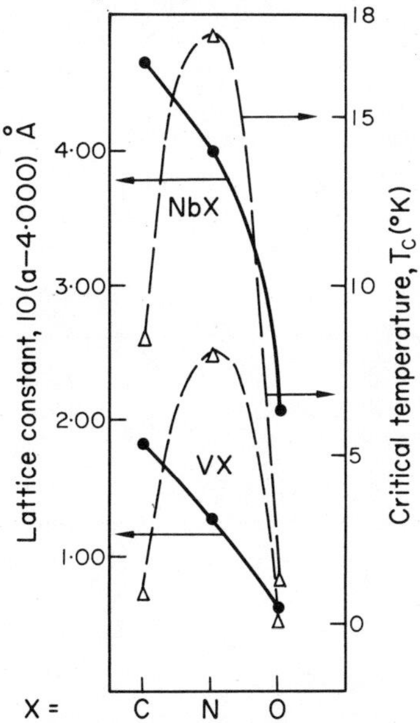

Fig. 19 Relation of high critical temperature to lattice constant. Maximum values of T_c occur with X = N (Nitrogen), while there is a bowing of the lattice constant with X = C, N and 0. (From Phillips, J. C., Physical Review Letters, **26.**)

used by Matthias, but rather the bond length between atoms in the crystal structure. The compounds RX^M with R = Ti, V, Zr or Nb and $X^4 = C$, $X^5 = N$ and $X^6 = 0$ all have the NaCl crystal structure. Some of these compounds have the same electron-atom ratio but have widely differing transition temperatures, so that this parameter cannot solely determine high values of T_c. Figure 19 shows that of these NbN (niobium

* Phillips, J. C., *Physical Review Letters* **26,** p. 543 (1971).

nitride) has a maximum critical temperature twice as large as VN (vanadium nitride), its lattice constant being also twice as large. The effect of an abnormally large lattice constant appears to be to stabilise the lattice vibrations so that a large attractive force between conduction electrons is achieved.

Table II

Calculated and observed maximum values of the critical temperatures for classes of similar materials.

METAL CLASS	T_c	T_c MAX CALCULATED	T_c OBSERVED	MAXIMUM MATERIAL
Pb	7·2	9·2	8·8	Pb-Bi
Nb	9·2	22	10·8	Zr-Nb
V_3Si	17·0	40		
Nb_3Sn	18·0	28	20·5	$Nb_{12}Al_3Ge$

4. Enhancement of Superconductivity

Probably the most challenging problem in superconductivity is the search for materials which will remain superconducting to temperatures well into the liquid hydrogen temperature range. The technological significance of materials which can pass currents without electrical loss has made this search one which has been pursued in many laboratories. The problem is in essence that of finding attractive interactions between electrons other than phonons. Many mechanisms have been suggested, these including virtual excitation of excitions (bound electron-hole pairs), excitation of electrons in the f-level and excitation of low energy electronic states in the side chains of organic polymers. It is useful to consider means which have been proved useful already in enhancing superconductivity, and those which have been proposed as being useful avenues for research.

The empirical rule for the occurrence of superconductivity established by Matthias, that alloys or compounds with average numbers of valence electrons per atom on the low side of 5 and 7 will probably have high superconducting transition temperature, has led directly to the compound which has the highest critical temperature. This is a solid solution, $Nb_{12}Al_3Ge$, crystallising with the beta-tungsten structure discovered by Matthias to remain superconducting to 20·8 K. Alloys and compounds which appear to fit this empirical rule may nevertheless require stringent control during preparation, for the mechanical state, homogeneity and the presence of impurity atoms are all capable of controlling the final transition. Well annealed specimens show sharper and higher transitions as shown by a Russian team led by Alekseevskii* working on a compound, $Nb_{15}Al_4Ge_2$, closely related to that of Matthias. Figure 20 shows the importance of the annealing process, a maximum in the transition temperature being achieved by annealing at 700 C for some days. At present, the material investigated by Matthias remains superconducting to 20·8 K and can sustain 200 kilogauss at 14 K and an ultimate critical field of 400 kilogauss at 0 K. This material is, however, extremely brittle and considerable technical progress will have to be made before it can become commercially available as wire or tape.

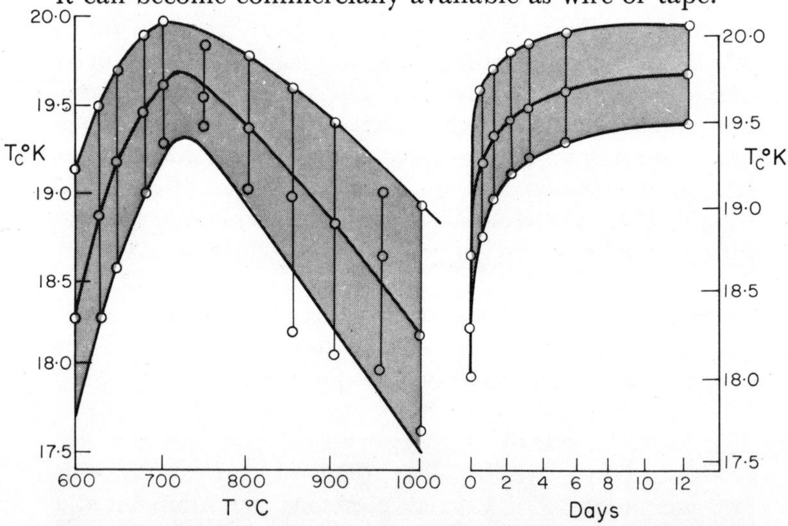

Fig. 20 Critical temperature of $Nb_{15}Al_4Ge_2$ *is increased by annealing at the correct temperature (left hand graph) and for the requisite time (right hand graph). (After Alekseevskii, N. E., Ageev, N. V., Mikhailov, N. N. and Shamrai, V. F.,* JETP Letters, **9,** *p. 16 (1969)*

There is a further way in which superconductivity has been enhanced with surprising ease. Layered structures of vacuum-evaporated superconducting and insulating

* Alekseevskii, N. E., Ageev, N. V., Mikhailov N. N., and Shainrai, V. F., *Soviet Physics JETP letters,* **9,** p. 16 (1969).

material, or of alternate layers of different metals, show considerably increased transition temperatures. It has been possible, for instance, to increase the critical temperature of bulk aluminium from 1·2 K to about 5·5 K using alternate 1 nm thick layers of aluminium with 1 to 2 nm layers of tin. This is a significant rise, although not in the high field superconducting range. The mechanism by which it is brought about is fairly well understood.

Such layered films are evaporated on to a substrate at low temperatures in order to obtain a completely clean film *in situ.* In this way the film comprises many small metallic crystallites separated by voids. At the surface, the fact that atoms share fewer atomic bonds leads to a lowering of the average frequency of the lattice phonons. This reduction can be accentuated when a new material is deposited on the first, being the outcome of an acoustic mismatch between the two lattices.

The behaviour of thin films has been considered by McMillan* who has emphasised that the behaviour of the lattice in superconductors is not described by the one parameter, the Debye temperature. The average value of the lattice vibrational frequencies is most important, and should this be lowered (in other words the lattice becomes softer) the superconducting transition temperature will rise. On this basis, to explain the enhancement of the critical temperature in films of Al and Sn, the average phonon frequencies should be decreased by about 30 per cent in the first few atomic layers of Al or Sn. Should the layers be made too thick, the average phonon frequency approaches the bulk metal value and the enhancement begins to decrease.

McMillan has calculated a maximum value from these arguments for the superconducting transition temperature

* McMillan, W. C. *Physical Review,* **167,** p. 331 (1968).

from known values of physical parameters for classes of 'similar" materials. These are given in Table II; vanadium-silicon alloys appear to be the most promising for further investigation.

A further possibility in the search for enhanced superconductivity exists in semiconductor materials. Normal semiconductors should not be superconductors because at low temperature the number of free charge carriers becomes extremely small, being governed only by the number of impurity atoms present. Germanium and silicon, two well known semiconductors, have been found to be superconducting, but only when subjected to high pressures. At 120 K bar pressure the critical temperatures are 5 K for germanium and 7 K for silicon.

The type of semiconductor which might be most suited to produce a technically useful superconductor has a many valleyed band structure—energy versus momentum. If such semiconductors are heavily doped, the number of electrons which populate these valleys becomes high. The existence of valleys in the band structure implies an increase in the coupling between electrons and lattice phonons, and thereby a greater possibility of electron-pairing mechanism which might lead to superconductivity according to the BCS microscopic theory. A high effective mass of the electron is of benefit, for it adds to the required electron-phonon interaction. As semiconductors have a high dielectric constant the chance of an attractive pairing between electrons is raised as the coulomb repulsion between electrons is thereby reduced.

Experimental tests of these arguments have been disappointing so far. The transition temperatures of materials like strontium titanate ($SrTi0_3$) is low, 0·1–0·4 K. In this particular material the critical temperature has a maximum of 0·3 K when doped to give 9×10^{19} charge carriers per cubic centimetre. This concentration is

100–1000 times higher than that used in semiconductors used in transistors. There are many possibilities open to change in semiconductors, and it is possible to independently alter charge carrier concentration, and the band structure, so that further research may be worthwhile.

Obviously the most important technical parameter of a superconductor is the critical current J_c as a function of magnetic field, no matter what the critical temperature. It is this parameter which can be altered fairly readily in known Type II superconductors to advantage, thereby enhancing the usable working range of the material. The way J_c depends upon the magnetic field (Figs. 3, 14), can be altered by treatment of the materials which introduce pinning sites for flux lines which penetrate the bulk material in the mixed state. Although such treatments do not in general result in a significant change in the value of the upper critical field, H_{c2}, the critical current is increased between H_{c1} and H_{c2} (Fig. 21). In this way the generation of heat through flux flow or flux jumps is diminished at a particular current required, say, in the design in a high field magnet.

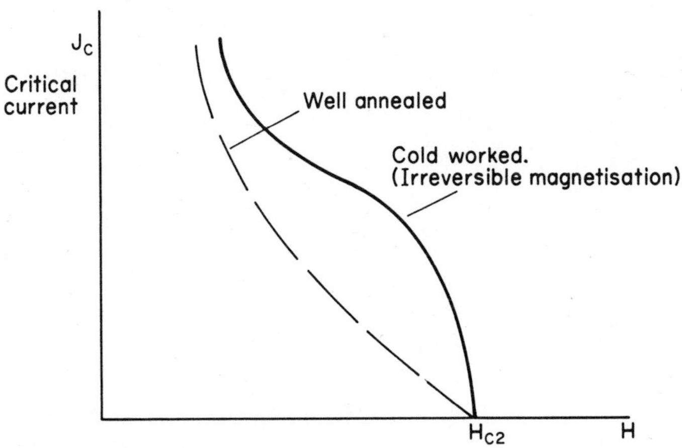

Fig. 21 Critical current increased by including pinning centres in the material

Treatments which produce this effect are generally those of introducing crystalline defects through mechanical deformation, rapid quenching, or cold working. Research with ion beams has led to the suggestion that small impurity particles in the 1–10 μm size range can be precipitated in the surface of a superconductor following ion implantation from bombardment. Such precipitates would act in a controlled fashion in altering the critical current.

This discussion of the enhancement of superconductivity would not be complete without mention of the proposal made in 1964 by Little* for a practical superconductor at room temperature. This proposal has been criticised however on theoretical grounds from many standpoints and there seems little prospect of it meeting with success.

Little suggested that certain organic polymers synthesised to a precise chemical specification would show zero resistance to the flow of charge along the length of the molecular chain. Built around large spine molecules of carbon atoms connected by alternate single and double bonds, the proposal called for side chains of dye molecules spaced along the spine at intervals.

The attractive pairing between electrons would come about as follows: one electron moving along the spine would polarise a side chain on passing it. A second electron would then be attracted to the region of positive charge at the near end of the side chain and so, indirectly, it would be attracted to the first electron. The choice of the chain molecules and the side chains would be important in making the attraction as large as possible, for the effect of attraction would be cumulative only if the movement of electrons along the spine was in harmony

* Little W. A., *Physical Review*, **134,** p. 1416 (1964). See also Reading List, Review Article 6.

with the movement of induced charge on the side chains. Two restrictions would be required also: only side chains associated with the spine should play a part in the mechanism, and the effect of one side chain should be independent of others.

The calculated critical temperature of such a molecule is well above room temperature, about 1,800 C, but the problem of synthesising a molecule to such a precise specification would be formidable. The proposal has been criticised from several points of view, and it has been shown conclusively that attractive pairing in this type of molecule would not take place because of the over-riding effect of coulomb repulsion between electrons, and fluctuations in the electron density. More fundamental than this is the criticism that the constraint of one dimensional movement can only hinder the required BCS type of attraction. Nevertheless, the advantages inherent in a room temperature superconductor are so great that this direction of thought will not be closed. At the present time an increase in critical temperature will most probably result from metallurgical developments aiming at compounds of some transition elements with a number of valence electrons per atom close to 5 or 7.

5. Superconducting Materials used in Technology

Apart from the searches for enhanced behaviour in superconductors, a great amount of technical development has taken place with certain known high field superconductors, and reliable superconducting wire and cable is available in many sizes. Most technical attention has been paid to wire, for this is the means by which magnetic fields are produced, and up to the present, the production of large magnetic fields without the need for continuous dissipation of large amounts of power has been the technological main use for the superconductor. Development of a special cable has been necessary, however, in order to provide stable operation of magnets to designed specifications. Such stable superconducting cables are either a single superconducting wire coated with a pure copper sheath, used at low fields, or an array of small wires embedded in a pure copper matrix for high field use.

The three superconducting alloys in most general supply are niobium-zirconium, niobium-titanium and niobium-tin (Table I). In the production of wire the first two, being ductile, can be drawn; niobium-tin, however, is extremely brittle and it must be prepared onto a suitable substrate so that it may later be wound into coil form.

In the production of niobium-zirconium and niobium-titanium, purity at all stages of the manufacturing process is important; impurities such as oxygen, nitrogen and carbon which go to interstitial sites and greatly affect the workability of the material must be excluded. Niobium is of the highest purity, being arc beam melted, and in producing the alloy the mixture is again arc beam melted under high vacuum. Wire is drawn from a bar or rolled to form strip. Final heat treatments for each type improve the current carrying capacity by producing the correct amount of defects for flux pinning. Of the two, niobium-titanium can be worked better; in this respect it resembles copper.

As niobium-tin is a brittle material, the usual means of drawing wire cannot be used, and it is produced as tape or wire by chemical techniques. Two mthods have been developed which can be employed in the continuous production of long lengths of wire.

In the vapour-deposition technique a continuous reduction by hydrogen of gaseous niobium and tin chlorides takes place, with a final deposition of niobium-tin alloys onto a stainless steel ribbon which acts as a substrate. The mixing takes place at between 700 and 900 C and the final reaction on the surface of the ribbon which is electrically heated to 1000 C.

Wires and tapes can be produced by diffusion, by coating a niobium wire or strip with a layer of tin formed by dipping, evaporation or electrodesposition. The whole is later heat treated at 1000 C to form Nb_3Sn by diffusion and a reaction at the boundary between the niobium and tin. In the case of wires produced in this way, mechanical strength is gained by forming a multistrand cable embedded in copper, the whole cable being heated to form niobium-tin.

SUPERCONDUCTING CABLES

As one of the main applications of superconductors is the production of large magnetic fields, the design of superconducting wire or cable used in long length in magnet construction must be suitable for reliable operation. Consideration of the mechanical strength of superconducting cables is essential, because the stresses on wires forming a solenoid can be considerable. These forces are due to the Lorentz force between current and magnetic field, and lead to a bursting force which is counteracted by a tension in the windings. An axial compressive force, which forces the turns of a solenoid together along the length of the coil, is also manifest.

These forces can be large in even reasonably sized magnets: in a typical 50 kG, 50 cm bore coil the total outwards force would be 3000 tons, resulting in a tension of 0·3 tons/cm^2; for larger magnets this tension may be 7–8 times higher. As mentioned earlier, to counteract the tendency for a magnet to prematurely revert to its normal state because of flux motion, large amounts of pure copper are incorporated in the cable to provide stabilisation. In this way the strength of the cable is about that of copper, 2 tons/cm^2. One way of increasing the cable's strength, therefore, is to harden the copper by cold working. A 10 per cent cold reduction in the cable during manufacture increases the proof stress at low temperature by four times. For coils which produce a bursting force greater than this figure, additional strengthening is required.

Axial compression must be balanced also by an adequate coil construction. In split coils this same force is present acting to pull the two halves of the coil together. The overall attractive force in a large split coil magnet may be several thousands of tons (e.g. in a bubble chamber magnet) and the structure must be able to withstand this.

The aspect of mechanical strength is further complicated by the requirement in many coils that liquid helium must be able to penetrate through at least some of the windings to ensure adequate cooling. Such an open structure will require further strengthening. The use of stabilised cables has simplified this requirement, but in magnets with high current densities, the requirement may still exist. Other means of cooling by use of channelled supercritical helium are being considered.

Although the design of small and large magnets is a specialist subject, with a considerable technical background, the aspect of the stabilisation of cables used in their construction is of fundamental importance. The development of stabilised superconducting cables has made large low temperature electrical engineering applications of superconductors possible. Such cables are designed to carry current densities to specified limits without degradation of performance. Their success means that both large and small magnets are now built with some form of stabilised cable.

The requirements for a stabilised cable stems from the degradation of the superconducting properties by the generation of heat within a Type II superconductor through flux jumping in the mixed state. Undesirably large local rises in temperature can arise, thereby nucleating a normal region which grows under Joule heating of the transport current. This effect will ultimately drive the whole superconductor normal at an unpredictable, prematurely low current.

When superconducting wires are embedded in a normal conductor of high conductivity (usually copper) any region of the superconductor which becomes normal by flux jumping does not spread. The copper provides a good thermal conduction path for heat to spread to the liquid helium reservoir so that the superconductor is quickly cooled to below its critical temperature and full superconductivity is restored. The copper acts also as a low resistance shunt across the normal region and current is shared between the two materials; after superconduction is restored the current returns fully to the superconductor. In practice, composite cables must be made with an adequate amount of copper to provide both a low resistance shunt path, and an adequate surface area for efficient cooling of the cable. The main cooling mechanism of a cable in the current shared mode is nucleate boiling of liquid helium at its surface, a type of convection cooling. Above a certain current, however, the heating rate becomes too great and film boiling sets in. Heat removal is then reduced by a factor of ten, or more, and the temperature of the cable increases with further degradation of the cable and a reversion to fully resistive behaviour.

Composite cables can be made, therefore, fully or partially stable (Fig. 22). In the fully stabilised case the copper surrounding the superconductor can carry current up to the critical current at the bath temperature without heating. In this case, any current displaced from the superconductor can be supported by the copper and any transient normal zone must collapse spontaneously (Fig. 22A).

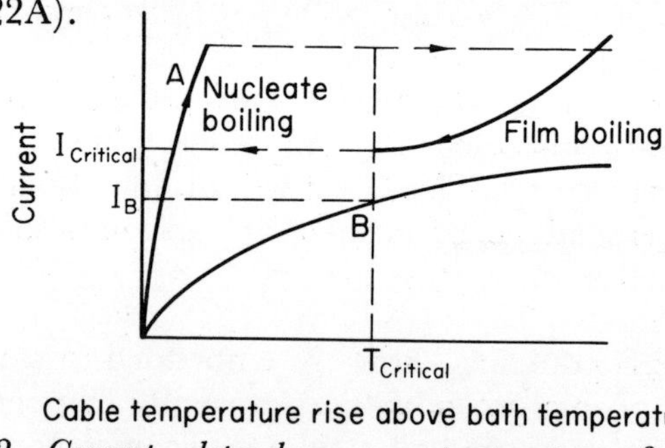

Fig. 22 Current dependence on temperature for composite cables, with different mechanisms for removing heat generated. A—fully stabilised cable up to $I_{critical}$ *of superconductor. B—Partially stabilised cable up to* I_B.

In small magnets the maximum operating current is designed to be well below critical. The amount of copper added is limited to that needed for stabilisation up to the design current. This constitutes partial stabilisation (Fig. 22B).

Composite cables contain many superconducting wires in order to obtain the required current capacity. The individual design of a cable depends upon the size and shape of the magnetic field to be produced, upon the means of cooling and upon the method of containment of the magnetic forces on the windings of a coil. Examples of composite cables are shown in Figs. 23 and 24. The best performance is obtained when superconducting filaments and copper are extruded together; the bond between the two is important and this process gives a good metallurgical-type diffusion bond.

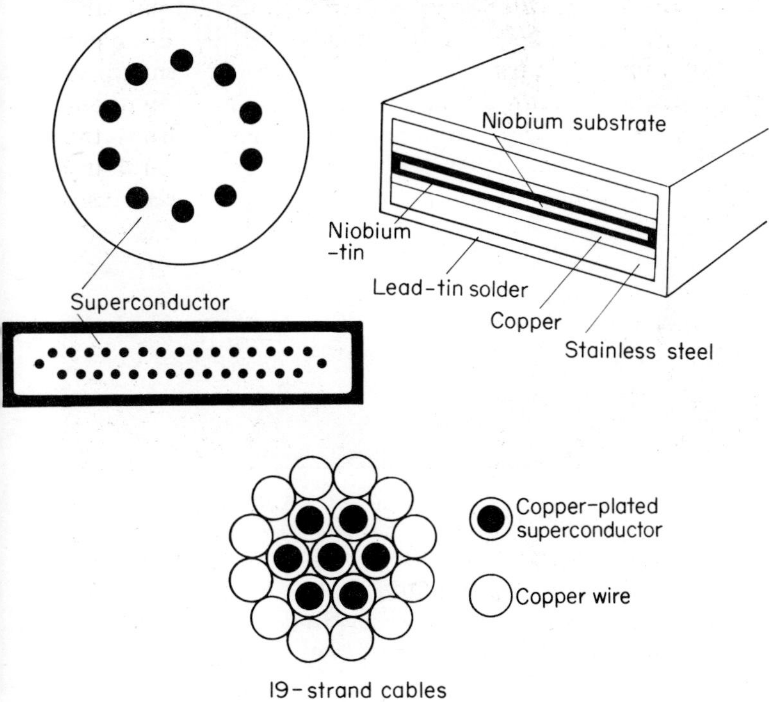

Fig. 23 Examples of stable superconducting cables

A further refinement in design gives the intrinsically stable superconducting cable. With many fine filamentary wires embedded in the copper matrix (Figs. 23, 24), the copper itself can support magnetisation current loops which span the width of the cable when current is changing. In this situation the effect of a flux jump in one strand can be communicated to other filaments as they are connected by low conductivity material. The energy of a flux jump is then no longer determined by the individual filament but by the whole width of the cable, depending upon the square of the cable's width.

Fig. 24 Photograph of the section of a superconducting cable using Niomax FM A61. The conductor uses twelve strands of superconductor (diameter 0·6 mm) over a copper core. The conductor is used in a pair of quadrupole magnets at CEA Saclay France. (Photograph by courtesy of Imperial Metal Industries (Kynoch) Ltd., New Metals Division, Birmingham)

Such magnetisation loops have a finite length, however, which depends upon the rate of change of the magnetic field. By twisting the cable to a pitch less than this length, the magnetisation currents can be suppressed and the cable remains intrinsically stable up to a particular rate of field change. When radial or axial field gradients are appreciable the filaments should be transposed rather than simply twisted. Higher rates of change of field, say in pulsed magnets, can be accommodated if the cable matrix has a lower conductivity than copper, say by using cupro-nickel. Intrinsically stable superconducting cables are required only in the largest magnets where, without their use, the charging time for the coil would be excessively long, and in magnets which are designed to be pulsed on and off.

6. Uses of Superconductivity

LARGE MAGNETS

The most important use of superconductivity is in the production of large magnetic fields over appreciable volumes without a large consumption of electrical power. In this area of use are superconducting solenoids which can produce 100 k gauss continuously in bores of about 1 inch, and larger volume magnets, for instance similar to that constructed at the Argonne National Laboratory with a diameter of 16 ft and height 10 ft producing 20 kG (Fig. 25).*

The use of superconducting magnets is not confined to one area of application, but at present the greatest development is taking place in high-energy physics. Associated with large particle accelerators such as the proposed European 300 *GeV* proton synchrotron are many magnets of high field and high homogeneity for bending and guiding the accelerated particle beam. Magnets are required in the accelerator ring itself and in the experimental target area. Such magnets are relatively long, 2 m, and have apertures between 10–20 cm and using conventional conductors require a continuous power of 30–100 kW for operation. Associated with the accelerator are bubble and spark chambers for particle detection, both of which require magnetic fields over large volumes.

It has been proven that superconducting magnets can be fabricated that would replace adequately conventional iron cored magnets with economies in both capital and running costs. For very large bubble chambers the power dissipation in a conventional design is very high, about

* See Reading List, Reports 2—Applied Superconductivity Conferences, 1969.

Fig. 25 16 ft superconducting magnet made at the Argonne National Laboratory. (Courtesy of U.S. Information Service)

11 MW for the type of magnet shown in Fig. 25, whereas using superconducting cables the energy required is little more than that finally stored in the magnetic field, and in the persistent mode only the refrigeration power to keep the magnet running.

The use of intrinsically stable cables shows also a substantial reduction in size and greater versatility in the geometry of the windings, since no space is required for the circulation of coolant. Stability in the persistent mode is excellent and designs can bring the homogeneity to 1 part in 10^6 over 1 cm^3 in small sized magnets.

High field magnets of large dimensions are used in a related field of research, that is the generation of electricity by magnetohydrodynamics (MDH), and in the containment of plasmas in work associated with attaining controlled nuclear fusion. In such work, the optimum magnetic field configuration can be complicated so that the flexibility in design of coils which comes by using superconducting cables commends this means of producing large volume fields to these areas of research.

On the industrial side, the possibility of using large magnetic fields without power loss has at last found expression in one or two projects. The superconducting homopolar motor developed by the International Research and Development Company Ltd.* and now driving the main water pump at Fawley power station is the first large scale motor (Fig. 26). It has produced 3250 hp at its maximum scheduled speed of 200 rev/min. The armature of this motor runs in the 88 inch bore of the superconducting solenoid. The applications for such motors are those which require continuous service without any great changes in speed, such as steel rolling mills, ship propulsion and power station auxiliary motors.

* See Reading List, Reports 2. Applied Superconductivity Conference 1969.

The development of a.c. machinery, however, must await a solution to problem of heat generated by flux motion when a.c. currents are carried by superconductors. At present, these losses are too great to contemplate the use of superconductors in a.c. engineering applications.

Fig. 26 Superconducting motor generating 3250 bhp at 200 rev/min installed at Fawley Power Station to drive main water cooling pump. Motor constructed by International Research and Development Co. Ltd., Newcastle-upon-Tyne. (*Photo courtesy I.R.D. Ltd.*)

The fundamental property of exclusion of magnetic flux from a superconductor and the consequent levitation of superconducting bodies above magnetic materials to reduce sliding friction has not been used in any technological proposal. There is a proposal however, now being tested, to lift a train off its tracks in order to reduce friction and thereby increase the speed of the train. Persistent current magnets would be carried on the train which would be guided by aluminium rails. In moving, the magnetic field of the train would induce eddy currents in the aluminium rails which, according to Lenz's law would act to minimise the change in flux threading the aluminium. Although the currents in the rails suffer from resistive losses, should the train be travelling fast, a levitating force will be present of sufficient magnitude to lift the train free of the rails. The minimum speed appears to be about 50 m.p.h. for a reasonable size of train carrying fairly large field magnets. At 300 m.p.h. the drag would be almost entirely that due to air, the drag forces due to Joule heating in the rails decreasing inversely as the square root of the speed. Propulsion could be by linear induction motor, using possibly the same magnets in the train together with coils in the track.

Many proposals have been made for superconducting power transmission lines and these have provoked many arguments about their feasibility on technical and economic grounds. Basically the idea is sound, that extremely large amounts of d.c. power can be transmitted over large distances without loss. A single cable made of a high field superconductor refrigerated to 4 K would have the capacity to supply the whole electrical requirements of a large city. Model superconducting cables have been built and fairly large currents have been carried over short distances.

At present there are probably equal arguments in favour of and against superconducting cables. The relatively low running costs must be balanced against the large capital costs. On the technical side, the problem of achieving the required dielectric insulation strength in a low temperature cable is not yet solved. There are problems also in providing appropriate safeguards to ensure the maintenance of current when a fault develops in the transmission line, refrigerating plant, or cable. A fault in the low temperature shield of a superconducting cable could not be mended quickly. The proposals for such cables have nevertheless stimulated a critical appraisal of cable design in general. With the rapid growth in the demand for electrical power, larger amounts of power will have to be moved to urban areas. The advantages of loss-less transmission by superconductor are too great to be ignored.

One proposal which is being developed is the superconducting linear accelerator. This application for superconductivity does not call for magnetic fields, but it does promise to use large amounts of superconductor, probably high purity niobium, or lead. In linear accelerators, electrons are accelerated in an electromagnetic travelling wave produced in a series of microwave cavities fed by klystrons. The electric field in the cavities is limited mainly by the radio losses in the walls of the cavities, and in conventional accelerators the microwave energy is turned on for very short periods followed by a quiescent time in which heat is removed.

At the frequencies used the surface losses in superconducting lead or niobium are 510^5 times smaller than in copper at room temperature, and quality factors of 10^{11} for the microwave cavities have been reached at temperatures of 1·8 K, using superfluid helium as refrigerant. A usual quality factor for a copper cavity at room temperature is only 10^4 to 10^5. Further reduction in the radio frequency losses of the accelerator can be obtained by using a lower microwave frequency, so that by operating at 1·8 K a continuous-wave accelerator is feasible. The use of superfluid helium is an integral part of the design. Superfluid helium has the property of extremely large (near infinite) thermal conductivity.

Cooling is achieved so efficiently with superfluid helium that the several hundreds of watts of heat developed over the length of the accelerator dial not at any point alter the temperature of the structure by more than ±0·02 K. This temperature stability is a requirement of the design to ensure stable operation. Figure 27 shows the basic design of the superconducting linear accelerator.

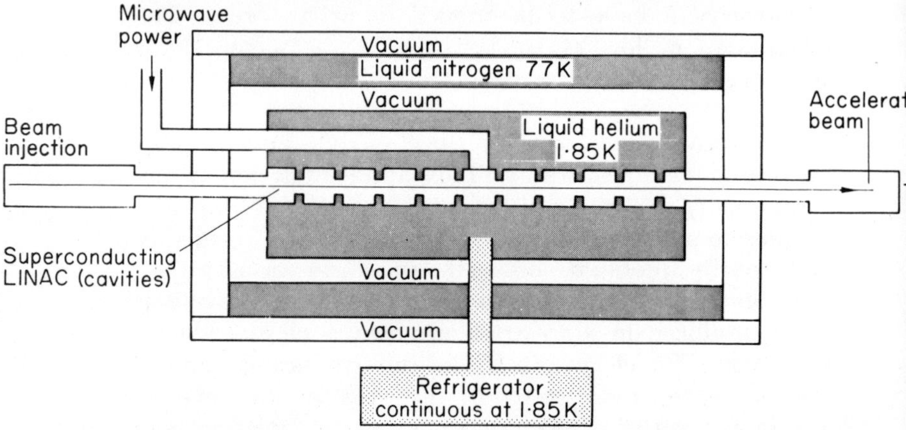

Fig. 27 Diagram of a superconducting linear accelerator cooled to 1·85 K

A section of the superconducting accelerator 1·5 m long has been built and electrons of 6 Mev have been obtained. This is to be extended to 160 m in length for further tests. These tests have been made possible by the development of a refrigerator operating at 1·8 K capable of removing 300 watts. Further development of the superconducting linear accelerator could lead to the replacement of the 2 mile long Stanford linear accelerator by a superconducting one, with a doubling of the energy of its output beam.

SENSITIVE SUPERCONDUCTING INSTRUMENTS

Superconductors have been used in the construction of delicate instruments designed to measure extremely small quantities. The amount of superconducting material used is usually small in comparison with magnets, but the zero resistance to the flow of current gives such devices an advantage over those of conventional design.

Persistent magnetic fields are highly stable for long periods of time, and this provides a means of levitating other superconducting bodies without friction. One device using this principle is the superconducting gyroscope in which a rotating superconducting sphere is supported in a carefully designed coil system which gives the sphere stable equilibrium. Although much progress has been made with this device it has not achieved adequate superiority over conventional gyroscopes using electrostatic levitation. Such low temperature gyroscopes have drift rates less than 10^{-2} seconds of arc per year, and the same angle can be read without disturbing the gyroscope.

The superconducting gravimeter in which a hollow sphere is levitated by superconducting coils has shown a better long term performance than conventional devices. Any change in the net acceleration in the system produces a change in the position of the sphere relative to the coils. The persistent nature of the magnetic fields and the exclusion of flux from the sphere produce a levitation system which is in principle perfectly stable. Drift rates of the device appear to be about 1 part in 10^9 over thirty days of observation. Very long period changes in the earth's gravitation due to earth and atmospheric tides should be detected by this device. The ultimate resolution is set by thermal fluctuation, and the finite penetration of magnetic flux into the surface of a superconductor.

The fact that a superconductor has zero resistance implies that the Johnson noise is zero from a superconducting device. In the measurement of very small voltages, therefore, any amplifier which uses superconductors in the first stage of amplification will achieve an enhanced signal-to-noise ratio. Two devices which use this property are the d.c. amplifier (Fig. 28), which is now produced commercially, and the magnetometer (Fig. 29).

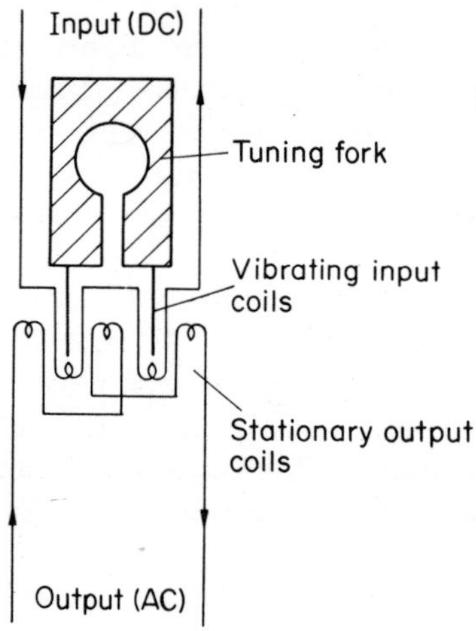

Fig. 28 Superconducting D.C. amplifier

In the d.c. amplifier, the current which flows through a small superconducting coil when attached to the voltage source to be measured, is detected by a second coil placed in close proximity. The first coil is attached to a tuning fork and vibrates so as to modulate the mutual inductance between the two coils. An a.c. signal is generated at the output, therefore, which is greater than the input voltage, and this can be further amplified by conventional means. The limit of resolution is set by the Johnson noise of the leads, and by any lack of constancy in the amplitude or frequency of the vibrating coil. D.C. voltages as small as 10^{-12} V can be detected, with a time constant for the amplifier of about 4 seconds.

In the magnetometer (Fig. 29) the magnetic field to be measured is presented to a zig-zag coil completed by a loop. The magnetic field creates a persistent current in the circuit. A superconducting sheet attached to a piezoelectric crystal placed close to the circuit alters the self inductance of the circuit when it is vibrated. Hence the current flowing in the circuit is modulated and this can be detected by a suitable superconducting pick up. Calculations show that the minimum magnetic field which can be detected is about 10^{-11} gauss.

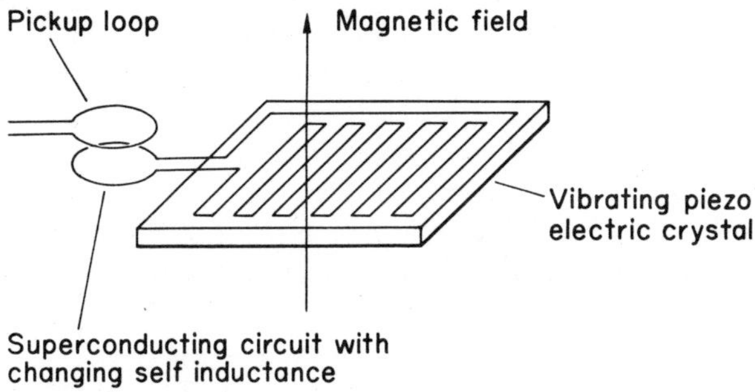

Fig. 29 Superconducting magnetometer

JOSEPHSON EFFECT AND DEVICES

A further set of superconducting devices depends upon the quantum mechanical wave nature of electrons in superconductors. Because of the attractive pairing of electrons the quantum mechanical waves describing every pair of electrons within a superconductor have the same phase and wavelength over the whole superconductor. So long as the material remains in this state it is not possible to introduce a phase difference in this wave between two parts of the specimen.

Considering two well separated pieces of superconductor the pair phases within each are the same, but the phases between the two bear no relationship. However, if the two are brought together, being separated by an insulating layer of about the same thickness as the coherence length of an electron pair, 100–200 nm, pairs can pass through the layer by quantum mechanical tunnelling.

During tunnelling no energy is expended to break up the pair, and because of the effect the phases of electron pairs in the two superconducting pieces become strictly related. This effect was theoretically and experimentally investigated by Josephson in 1962.* Electron pairs can be exchanged without any voltage appearing across the insulating layer, this being known as the d.c. Josephson effect. Above a critical current through the layer a voltage appears and the junction behaves resistively (Fig. 30). The energy gained by the pair from the junction voltage is emitted as electromagnetic radiation, its frequency being related to the energy by the Planck formula, $f = (2e/h)V$, where h is Planck's constant and V the junction voltage. This is known as the a.c. Josephson effect.

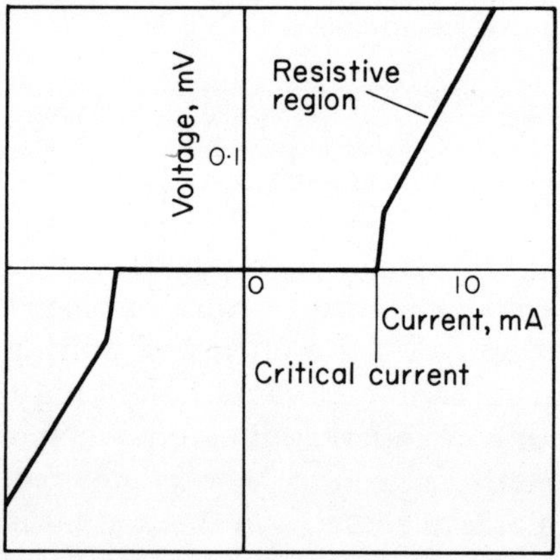

Fig. 30 Voltage versus current in a D.C. Josephson junction

Several forms of junction which exhibit the Josephson effect are shown in Fig. 31. In some, two evaporated films of niobium are used, the first being partially oxidised to form the insulating layer before the second is laid down. In others a point contact is used, the oxide layer

* Josephson, B. D., *Physics Letters*, **1**, p. 251 (1962) and reading list (Lagenberg).

across the point acting as the weak link between the two superconductors. In the a.c. Josephson case a double junction made of two connected in parallel is often used.

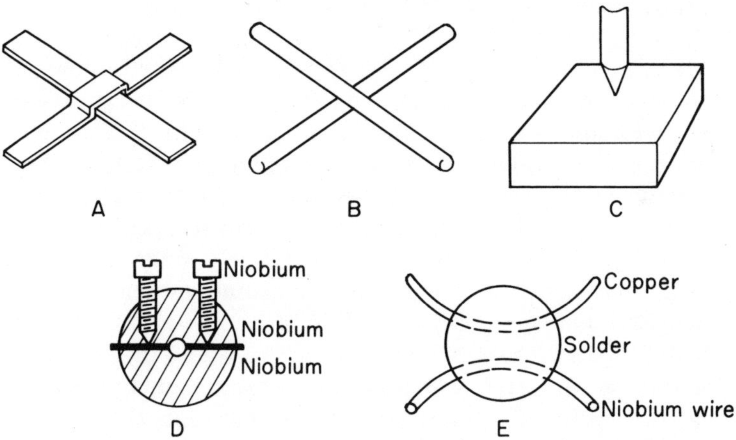

Fig. 31 Several forms of junction showing Josephson Effect. A, Evaporated film; B—C, Point contact; D—E, Point contact double junctions

Interference between the electron waves crossing each junction gives rise to peaks in the critical current which can flow through the double junction (Fig. 32). Peak currents occur for changes in magnetic flux threading the area surrounded by the double junction of one half of the value of the fundamental quantum of flux, $h/2e$ equal

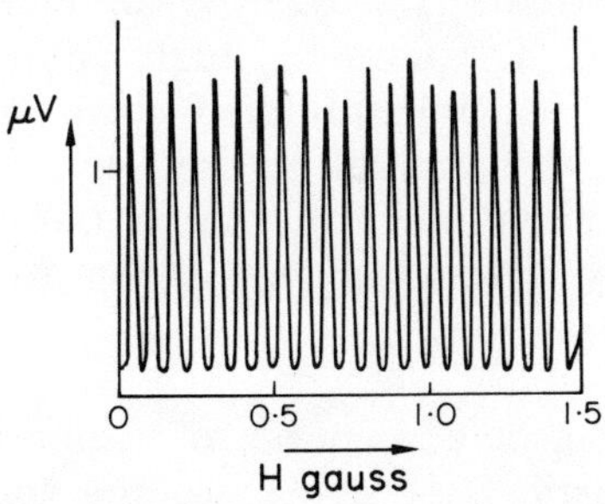

Fig. 32 Voltage oscillations in a tin-lead double junction

to $2{\cdot}07 \times 10^{-7}$ gauss/cm^2, or $2{\cdot}07 \times 10^{-5}$ microtesla. The double junction acts very much in the same way as an optical interferometer in which interference of two beams of light is detected.

Many new superconductive devices based on the Josephson effect have been constructed. These have been based preferentially on the a.c., or interference effect, although the d.c. effect gives rise to a very efficient switch. A d.c. Josephson junction "gate" can be switched by a nearby "control line from its superconducting to resistive state under the influence of the control line's magnetic field. With two switches connected in parallel, current can be switched between the two. The state of the combined switch is stored in the Josephson superconducting current of one or the other junction. Computer memories based on this principle have been proposed, for the switch promises to be much faster than other types of superconducting switches which have been suggested in the past.

In the a.c. effect is embodied a means of relating exactly a voltage to the frequency of radiation emitted from a junction. Radiation can also be absorbed, a current flowing across the junction when the junction is biased to the correct voltage. Microwaves at 10 GHz have been emitted from a junction mounted in a low temperature resonant cavity. Powers of 10^{-11} W are emitted; although small, these are adequate for high accuracy frequency measurement. Such measurement gives the value of the voltage biasing the junction to the same accuracy, about 1 part in 10^{10}. Such inherently simple but highly accurate voltmeters have been suggested for use in national standards laboratories.

In the detection, or absorption, of radiation a Josephson junction is particularly suited to the far infrared. By biasing a junction to a suitable operating voltage, the equivalent of 5×10^{-13} W noise power can be detected. The response time is less than 10 ns. At lower frequencies, in the hundreds of megahertz, junctions can be utilised to detect the absorption of radio waves by nuclear magnetic resonance. With cobalt, detection of 10^{-16} W absorbed in a total irradiation power of 4×10^{-12} W has been possible, making this type of superconducting device into an ultra-low power spectrometer of considerable operating bandwidth.

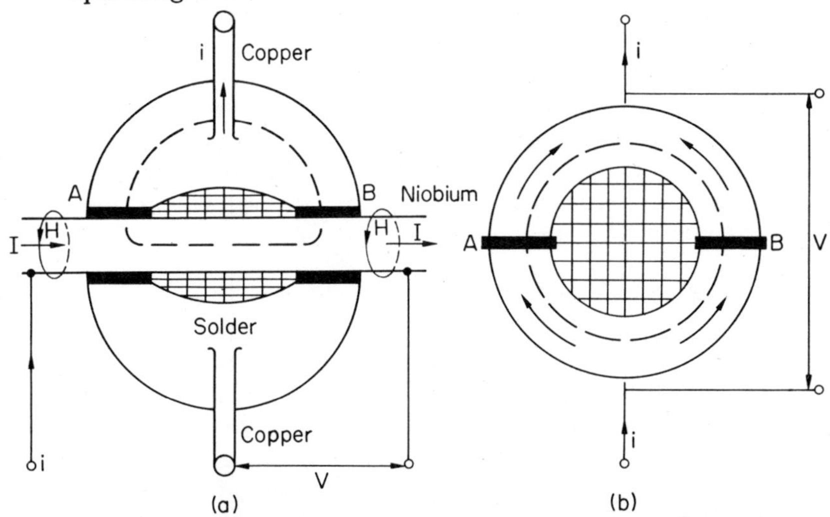

Fig. 33 Simple double junction made with a blob of solder round a niobium wire (a). This acts in a similar fashion to the arrangement (b) with junctions at A and B. Current (i) entering divides through A and B to enclose the hatched area through which magnetic flux passes

In double junctions, the interference effect between pairs of electrons makes for considerable resolution of magnetic field, or current. The simple double junction made with solder enclosing niobium wire (Fig. 31E and Fig. 33) has been used to detect voltages down to 10^{-14} V with a time constant of 1 sec. Any current flowing in the niobium wire is sensed by the effect its accompanying magnetic field has on the critical current of the interference junction between niobium wire and solder. A current of 1 μA can be detected by sensing the voltage between niobium and solder. As the device has negligible resistance, say 10^{-8} ohm, and a very low inherent inductance, about 10^{-8} henry, the sensitivity as a voltmeter is good. When used as a low level null detector, 10^{-12} ohm can be measured. The periodic changes in critical current of an interference device can be counted as an external magnetic field is changed. The variation in magnetic flux is therefore detected as a direct multiple of the fundamental quantum of flux $2{\cdot}77 \times 10^{-7}$ gauss/cm^2, and the device can act as a magnetometer. In a commercially made magnetometer using this principle, special add and subtract circuits count the steps and enable both increasing and decreasing fields to be measured. The range is 3000 gauss with field changes of 10^{-7} gauss detected.

EPILOGUE

In principle many superconducting devices have sensitivities always greater than conventional techniques. The simplicity which accompanies some devices commends them for long term service. Clear technological progress in delicate sensitive devices will be made, and these will become an accepted part of future laboratory measuring technique.

In the big league of superconductors, the proven usefulness of large magnetic fields maintained without significant power consumption ensures that the use of superconductors will be widespread in future technology. Their use will have to be tempered by adequate tests to prove their reliability for continuous operation, and their economic advantage. The much searched for high temperature superconductor may not appear for many years, but should it ever do so electrical engineering will rapidly change in outlook to accommodate its discovery.

READING LIST

TEXTS

1. Shoenberg, D. "Superconductivity", Cambridge University Press (1965).
2. De Gennes, P. G. "Superconductivity of Metals and Alloys," Benjamin Inc. (1966).
3. Newhouse, V. L. "Applied Superconductivity", Wiley and Sons (1964).
4. Rose-Innes, A. C. and Rhoderick, E. H. "Introduction to Superconductivity," Pergamon (1969).
5. Fishlock, D. "A Guide to Superconductivity," MacDonald Elsenier (1969).
6. Kropschot, R. H., Birmingham, B. W. and Mann, D. B. "Technology of Liquid Helium," NSB Monograph III, Washington (1968).
7. Mendelssohn, K. "The Quest for Absolute Zero," Weidenfeld and Nicolson (1966).

REVIEW ARTICLES

1 Gorter, C. J. Editor. Various topics in superconductivity, *Progress in Low Temperature Physics*, **1–6,** (North-Holland Publishing Co.).
2. Matthais, B. T. The Empirical Approach to Superconductivity, *Advances in Cryoengineering*, **13,** p. 1. (1969) (Plenum Press).
3. Bardeen, John, Advances in Superconductivity, *Physics Today*, October, 1969.
4. Goodkind, John M. Applications of Superconductivity, *American Journal of Physics*, **39,** p. 9. (1971).
5. Ginsburg, N. I. The Problem of High Temperature Superconductivity, *Contemporary Physics*, **9,** p. 355 (1968).
6. Little, W. A. Superconductivity at Room Temperature, *Scientific American*, **212,** p. 21 (February, 1965).
7. Brandt, N. B. and Ginsburg, N. I. Superconductivity at High Pressures, *Contemporary Physics*, **10,** p. 355 (1969).

8. Brandt, N. B. and Ginsburg, N. I. Superconductivity at High Pressure, *Scientific American*, **224,** p. 83 (April 1971).
9. Sampson, W. B., Craig, P. P. and Strong, Myran. Advances in Superconducting Magnets, *Scientific American*, **216,** p. 114 (March, 1967).
10. Langenberg, D. N., Scalapino, D. J. and Taylor, B. N. The Josephson Effect, *Scientific American*, **214,** p. 30 (May, 1966).
11. Taylor, B. N., Parker, W. H., Lagenberg, D. N. and Denenstein, A. On the use of the A.C. Josephson Effect to maintain standards of electromotive force, *Metrologia*, **3,** p. 89 (1967).

REPORTS

1. Superconductive Materials and some of their properties. B. W. Roberts General Electric Research Laboratory Reports 63–RL–3552M
   64–RL–3540M
   Also NBS Technical Note 408 (1966).
2. Applied Superconductivity Conference Reports. *Journal of Applied Physics* Vol. 39, No. 6, 1968
   Vol. 40, No. 5, 1969
   Vol. 42, No. 1, 1971.
3. International Conferences on Low Temperature Physics (Up to Twelfth held in 1970).
4. Proceedings of the International Conference on Organic Superconductors. (W. A. Little, Editor, Interscience 1970).

INDEX